Controlling Exposure - Diesel Emissions in Underground Mines

Controlling Exposure - Diesel Emissions in Underground Mines

Editor

Marcel Talbot

scitus academics

Controlling Exposure - Diesel Emissions in Underground Mines
Edited by **Marcel Talbot**

ISBN: 978-1-68117-454-9
Library of Congress Control Number: 2016937622

SCITUS Academics LLC,
www.scitusacademics.com
Box No. 4766, 616 Corporate Way,
Suite 2, Valley Cottage,
NY 10989

Printed in United States of America on Acid Free Paper

Preface

The exposure of mine workers to diesel particulate matter (DPM) and gases is an issue of great concern to the underground mining community. As industry works to achieve compliance with these standards, mine operators are looking for feasible methods for reducing DPM concentrations in their mines. In addition, the industry needs methods to accurately measure DPM to ensure that the control strategies they adopt are working successfully. The use of diesel-powered equipment by the underground mining community has continuously increased over the last several decades. A similar trend is seen in underground metal/nonmetal mines. This extensive utilization of diesel-powered equipment generates the potential for exposure of underground miners to particulate matter and gaseous emissions from this equipment and creates an important challenge for the mining industry worldwide to control these emissions. In recent years, studies have shown that long-term exposure to combustion-related fine particulate pollution, including diesel particulate matter (DPM), is an important risk factor for cardiopulmonary and lung cancer mortality. Therefore, health issues associated with exposure to DPM are receiving substantial attention from the public, government agencies, and academia. The book Controlling Exposure - Diesel Emissions in Underground Mines present an overview of the integrated approach and review the strategies being used by mines to reduce the concentrations of DPM emitted from the tailpipe, as well as present results from research evaluating these strategies.

Table of Contents

Chapter 1

Mutagenicity of Biodiesel or Diesel Exhaust Particles and the Effect of Engine Operating Conditions

Elena R Kisin[1], X.C Shi[2], Michael J Keane[2], Aleksandar B Bugarski[3] and Anna A Shvedova[1*]

[1]*Pathology and Physiology Research Branch, Health Effect Laboratory Division, National Institute for Occupational Safety and Health, Morgantown WV, USA.*

[2]*Exposure Assessment Branch, Health Effect Laboratory Division, National Institute for Occupational Safety and Health, Morgantown WV, USA.*

[3]*Office of Mine Safety and Health, National Institute for Occupational Safety and Health, Pittsburgh, PA, USA.*

ABSTRACT

Background: Changing the fuel supply from petroleum based ultra-low sulfur diesel (ULSD) to biodiesel and its blends is considered by many to be a viable option for controlling exposures to particulate material (PM). This is critical in the mining industry where approximately 28,000 underground miners are potentially exposed to relatively high concentrations of diesel particulate matter (DPM). This study was conducted to investigate the mutagenic potential of diesel engine emissions (DEE) from neat (B100) and blended (B50) soy-based fatty acid methyl ester (FAME) biodiesel in comparison with ULSD PM using different engine operating conditions and exhaust aftertreatment configurations.

Methods: The DPM samples were collected for engine equipped with either a standard muffler or a combination of the muffler and diesel oxidation catalytic converter (DOC) that was operated at four different steady-state modes. Bacterial gene mutation activity of DPM was tested on the organic solvent extracts using the Ames *Salmonella* assay.

Results: The results indicate that mutagenic activity of DPM was strongly affected by fuels, engine operating conditions, and exhaust aftertreatment systems. The mutagenicity was increased with the fraction of biodiesel in the fuel. While the mutagenic activity was observed in B50 and B100 samples collected from both light- and heavy-load operating conditions, the ULSD samples were mutagenic only at light-load conditions. The presence of DOC in the exhaust system resulted in the decreased mutagenicity when engine was fueled with B100 and B50 and operated at light-load conditions. This was not the case when engine was fueled with ULSD. Heavy-load operating condition in the presence of DOC resulted in a decrease of mutagenicity only when engine was fueled with B50, but not B100 or ULSD.

Conclusions: Therefore, the results indicate that DPM from neat or blended biodiesel has a higher mutagenic potency than that one of ULSD. Further research is needed to investigate the health effect of biodiesel as well as efficiency of DOC or other exhaust aftertreatment systems.

Keywords: Ames *Salmonella*/microsomal assay, mutagenic activity, biodiesel, diesel exhaust particulate matter, oxidation catalytic converter

BACKGROUND

Biodiesel is considered an attractive alternative to diesel oil since it can be produced by domestic natural sources, reducing dependence on petroleum-based fuels [1]. Biodiesel can be produced from different plant oils by transesterification of triglycerides from vegetable oils with ethanol [2,3]. The increasing production and consumption of biodiesel has encouraged researchers to assess its hazard and fate in the environment. Specifically, diesel combustion engines are an important component of the industrial and transportation sectors, including numerous mining, agriculture and construction uses. There are concerns, however, with exhaust emissions in regards to potential adverse health effects associated with exposure to diesel particulate matter (DPM). Diesel engine emissions are highly complex mixtures of aerosols and gases. They consist of a wide range of organic and inorganic compounds [4]. Most of the particles from diesel engine exhausts are of nano-scale and therefore readily respirable. These particles potentially have hundreds of chemicals absorbed onto their surfaces, including known and suspected mutagens and carcinogens, *e.g.*polycyclic hydrocarbons (PAH) and nitrated PAH (nPAH) [5]. The formation of PAH depends on the type of engine, fuel composition, the engine operating conditions and the effectiveness of exhaust aftertreatment [6]. Exposure to diesel engine emissions and their atmospheric transformation products occur often in both environmental and occupational settings. Compared to their parent PAH, most of the resulting compounds generated from the combustion system are mutagens or have an enhanced mutagenic potency [7-9]. A causal relationship of exposure to diesel engine emissions and lung cancer was suggestive for occupational settings but not for the general population [5]. According to two large studies [10,11] conducted among the nonmetal miners, diesel exhaust increases the risk of death from lung cancer. The International Agency for Research on Cancer [12], a part of the World Health Organization (WHO), classified diesel engine exhaust as carcinogenic to humans (Group 1) based on sufficient evidences to link exposure to an increased risk of lung cancer.

During recent years, strong efforts have been made to minimize diesel engine emission-related health hazards. This includes improved combustion, use of exhaust aftertreatment, the reduction of sulfur and aromatics content in fuels and lubricating oil, and the introduction of reformulated fuels [4,5]. Various diesel exhaust treatment technologies, such as diesel particulate filter (DPF) systems, disposable filters elements (DEDs), and diesel oxidation catalysts (DOCs) have been implemented. In mining, improvements of ventilation and the curtailment of diesel particulate matter (DPM) and toxic gaseous emissions from existing and new diesel powered equipment are commonly perceived as the most promising tools to meet Mine Safety and Health Administration (MSHA) regulations [4]. The use of biodiesel results in a substantial reduction of unburned hydrocarbons, carbon monoxide, and PM as compared to diesel emissions [13]. The issue of exposure to DPM and the use of biodiesel blends is particularly critical in the mining industry where approximately 28,000 U.S. underground miners are potentially exposed to relatively high concentrations of DPM [4]. Mine operators are currently using 25-100% biodiesel blends

[14,15]. The concentrations of biodiesel in the blends used in underground mining are substantially higher than those used in other on- and off-highway applications [4].

The effects of biodiesel on emissions were found to vary widely, in particular with usage conditions, engine type and age [16-18]. Mine studies showed potential of neat biodiesel [18] and biodiesel blends [19-23] to reduce exposure of underground miners to DPM. However, the combustion of biodiesel in diesel engines typically results in slight increase of nitrogen oxide emissions [24]. The particle-bound volatile organic fractions of DPM [17,25] were found to be higher for biodiesel than for petroleum diesel fuels [17,25]. This may impact the biologic effects and toxicity of biodiesel exhaust particles.

Previous studies have compared the mutagenicity of diesel and biodiesel exhausts using bacterial reverse mutation assay [26,27]. This assay is suited as a screening tool for hydrophobic mutagenic compounds or mixtures, such as PAH from diesel engine emissions [5]. There is conflicting and limited information available regarding the mutagenic potential of biodiesel emissions specific to fuel category, engine type, and operating conditions [28-39].

The purpose of our study is to investigate the effects of soy-based fatty acid methyl ester (FAME) biodiesel, engine operating conditions and exhaust aftertreatment on the mutagenicity of diesel aerosol organic solvent extracts in *Salmonella typhimurium*. The samples analyzed in this study were collected directly in an underground mining environment of the National Institute for Occupational Safety and Health (NIOSH) Lake Lynn Experimental Mine by using the NIOSH Diesel Laboratory.

METHODS

Sample collection

The DPM samples were collected at the diesel laboratory in the D-drift of the NIOSH Lake Lynn Experimental Mine. This laboratory was designed to allow evaluation of control technologies and strategies in an underground environment [23,40]. The tests were executed using Isuzu C240 engine rated at 41 kW, one of the most commonly used light-duty engine in US underground coal mines [41]. The engine was operated at four steady-state modes using eddy-current dynamometer (Figure 1).

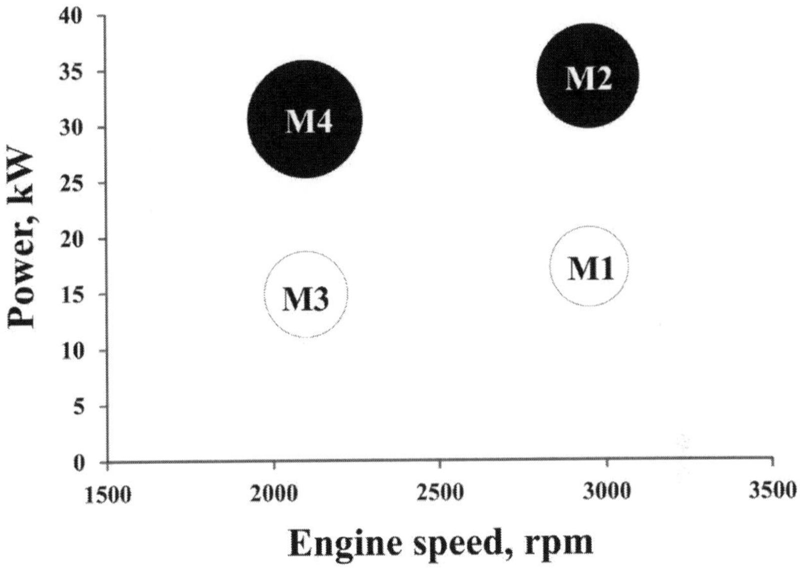

Figure 1. Engine operating conditions. Engine was operated at four steady-state modes. One of the selected modes at each of the engine speeds represented heavy engine load (M2 and M4 with an engine speed of 2950 rpm and 2100 rpm, torque of 108.4 Nm and 138 Nm, and power 34.3 kW and 30.6 kW, respectively), while the other was more representative of light engine load (M1 and M3 with an engine speed of 2950 rpm and 2100 rpm, torque of 63.7 Nm and 73.2 Nm, and power 17.2 kW and 14.9 kW, respectively).

These modes were selected to cover a wide range of engine operating parameters. One of the selected modes at each of the engine speeds was representative of heavy engine loads (M2 and M4), while the other two were representative of light or medium engine loads (M1 and M3). Two exhaust configurations (muffler and muffler plus DOC) were evaluated for M1 and M2 modes. Soy-based FAME biodiesel (B100) was supplied by Stepan Company (Stepansol SB-W, Bordentown, NJ). ULSD supplied by Guttman Oil was used as baseline fuel. The B50 blend was prepared at the site. The fractions of biodiesel and ULSD were determined volumetrically. The DPM samples were collected concurrently on 90 mm Teflon coated glass fiber filters (Pall TX40HI20-WW). The nominal sampling flow rate for each of the sampling stream was 91 lpm. The sampling times were between 120 and 180 min depending on the test. The flow rates were maintained using rotary vane pumps (Gast, Model 1023) and controlled with mass flow controllers (Sierra Instruments Inc., Model 840). The Teflon-coated aluminum cyclone (URG, Model URG-2000-30EH) was used to remove aerosols with D_{50}larger than 1 μm. The properties of sampled aerosols are published in [23]. Eighteen samples were collected for mutagenicity analysis. Table 1 shows the information on the fuel, the exhaust configuration and engine operating modes pertinent to each of the samples.

Table 1. Mutagenicity induced by DPM samples.

Exhaust Configuration	Engine Mode	Fuel	Revertants/plate[a] in concentration (μg/plate)			
			0.0[b]	13.3	40.0	120.0
Muffler	M1	ULSD	18±4.5	27±2.3	49±17.3	67±3.0*
Muffler	M2		18±4.5	23±5.0	30±0.8	53±7.9
Muffler	M3		18±4.5	24±3.4	40±5.9	71±4.3*
Muffler	M4		18±1.0	23±2.0	22±2.3	21±3.3
Muffler+DOC	M1		18±1.2	26±1.6	42±4.4	80±11.7*
Muffler+DOC	M2		18±1.2	24±1.2	33±4.8	42±7.7
Muffler	M1	B50	18±1.2	22±2.0	50±3.9*	79±1.5*
Muffler	M2		18±1.2	28±0.4	58±6.4*	71±4.1*
Muffler	M3		18±1.2	25±1.1	42±2.0*	80±9.9*
Muffler	M4		18±1.2	18±4.4	16±2.2	28±6.1
Muffler+DOC	M1		18±1.2	29±5.1	40±3.5	73±2.8*
Muffler+DOC	M2		18±1.2	29±0.6	37±5.3	60±6.2*
Muffler	M1	B100	18±4.5	33±13.1	63±7.2*	128±6.8*
Muffler	M2		23±0.6	55±0.6*	84±2.7*	148±25.7*
Muffler	M3		23±0.6	26±2.7	49±5.0	73±3.3*
Muffler	M4		27±1.1	24±4.9	21±1.1	35±6.0
Muffler+DOC	M1		18±4.5	19±6.1	26±2.1	40±15.2
Muffler+DOC	M2		32±1.1	30±1.1	87±3.8*	155±5.5*

[a] Mean value of two plates ± standard deviation.
[b] Solvent control was shared within each group of samples assayed simultaneously.
* Positive response.

Samples preparation

Each sample consisted of DPM collected concurrently on two 90 mm filters. The DPM were extracted simultaneously from both filters using 100 ml of acetone and sonicated for 2 hours. The extract was reduced by evaporation in N_2 environment and centrifuged at 4000 rpm for 40 minutes. The supernatant was removed and filtered through 25 mm sterile PTFE filter. The filtrate was transferred to a tared sterile brown vial and evaporated to dryness under N_2. Twelve milligrams of DPM extract per milliliter of dimethyl sulfoxide (DMSO) was prepared as a stock solution. Samples were diluted further with DMSO to obtain the required concentration.

Ames salmonella assay The mutagenicity of samples was determined using the preincubation variant of the Ames Salmonella assay system [42]. Results from the previous studies indicate that *Salmonella typhimurium*strains TA98 and YG1024 gave better mutagenic response than TA100 and YG1029 to DPM tested, and S9 microsomal activation was not necessary for examination of mutagenic activity in DPM [43-45]. Due to the limited amounts of the samples available, only *Salmonella typhimurium* TA 98 without S9 microsomal activation was used in the study. The cell suspension of *Salmonella typhimurium* TA98 was prepared by overnight incubation of a small piece of frozen culture in 25ml nutrient broth. The genotype of the bacteria was confirmed following the recommendation by [42]. A sample solution of 10µl, along with 65µl physiological saline, and 25µl of the TA98 cultures were pre-incubated at 37°C for 30 min on a rotary drum incubator before plating. The control was established using DMSO in place of sample solution. After incubation, 2.5ml of molten top agar were added to each sample tube, the contents mixed, and poured onto a Vogel-Bonner minimal media plate. The plates were incubated at 37°C for 48h prior to counting. The conditions of the bacterial background lawn were examined during every experiment in order to know if cytotoxicity appeared in the high concentrations. Three readings of the revertant colonies per plate were scored by an automatic colony counter (Accucount 1000, Biologics Inc., Gainsville, VA) after incubation. The average numbers of revertant colonies per each concentration were calculated. An extract was considered mutagenic if the number of revertants in any of the concentrations tested was two-fold or greater than the control [27] and showed a dose related response. All experiments were repeated twice to confirm the results.

Statistical analysis

Results were compared by One Way ANOVA using the All Pairwise Multiple Comparison Procedures (Holm-Sidak method). All results are presented as mean + standars error of the mean (SEM).

RESULTS

Effects of B50 or B100 in comparison to ULSD

The mutagenic effects of B50 and B100 were compared with those of the ULSD for all four tested engine conditions (Table 1, Figure 2). When engine was equipped with muffler, the concentration dependent mutagenic activity was observed for all samples collected at light-load (M1 and M3) and one of the heavy-load (M2) operating conditions. The mutagenic activity was increased with fraction of biodiesel in the fuel for all those 3 modes. B100 (120 µg/ plate) induce the strongest mutagenic effect at M1 and M2 (91% and 179% over ULSD, respectively). At the same conditions, B50 caused only 18% and 34% increase in mutagenicity over ULSD, respectively. B50 and B100 did not induce a significant increase (13% and 3%, respectively) in mutagenic activity when engine was operated under M3 conditions, as compared to ULSD. No positive mutagenic activity was detected for all samples collected at M4 (heavyload) operating conditions.

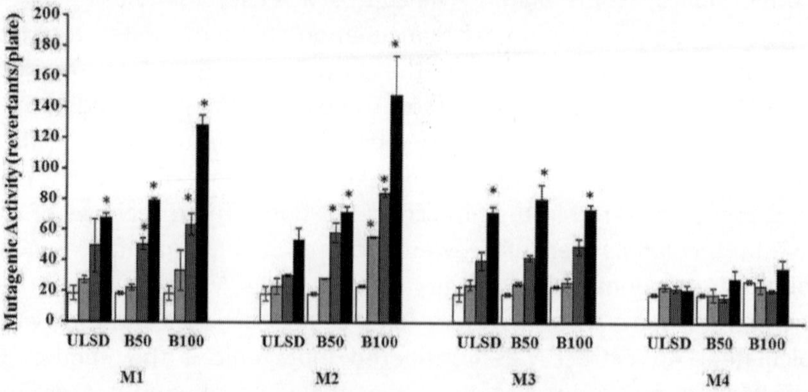

Figure 2. Mutagenic activity of DPM samples collected for the engine fueled with ULSD, B50 or B100 and equipped with muffler only at 4 steady-state modes. ULSD, B50 or B100 samples (0, 13.3, 40 or 120 µg/plate) were incubated at 37° C with the*Salmonella typhimurium* TA 98 without S9 microsomal activation. Clear columns – control samples (PBS exposure); light gray columns – exposure with 13.3 µg/plate of DPM; Dark gray columns – exposure with 40 µg/plate of DPM; Black columns – exposure with 120 µg/plate of DPM. Data represent mean values (+SEM) of the average number of revertant colonies per sample. Each sample was tested twice in two separate experiments. *positive responses, as evidenced by the number of revertant colonies being at least two-fold greater than the respective control value.

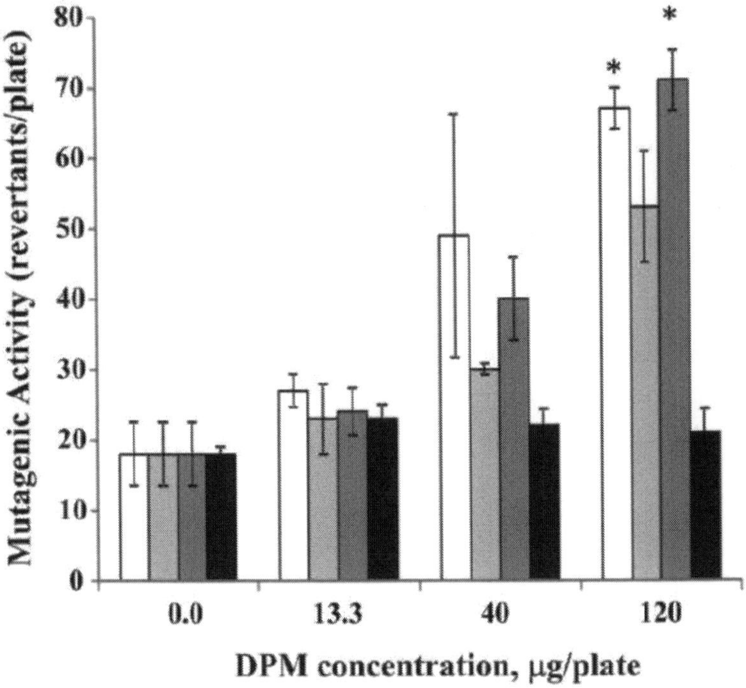

Figure 3. Effect of engine operating conditions on mutagenic activity of ULSD samples collected for the engine fitted with muffler. ULSD samples (0, 13.3, 40 or 120 µg/plate) were incubated at 37o C with the *Salmonella typhimurium* TA 98 without S9 microsomal activation. Clear columns – mode M1; light gray columns – mode M2; Dark gray columns – mode M3; Black columns – mode M4. Data represent mean values (+SEM) of the average number of revertant colonies per sample. Each sample was tested twice in two separate experiments. *positive responses, as evidenced by the number of revertant colonies being at least two-fold greater than the respective control value.

Effect of engine operating conditions on mutagenic activity of DPM

The effects of the four engine operating conditions were studied using the samples collected from engine equipped with a muffler (Table 1 and Figures 3,4 and 5). When engine was fueled with ULSD, the mutagenic activity was observed for the samples collected for light-load engine operating conditions (M1 and M3), but not for heavyload engine operating conditions (M2 and M4), and only for the highest studied concentration of 120 μg/plate (Figure 3). When engine was fueled with B50 (Figure 4) and B100 (Figure 5), the mutagenic activity was detected at both light-load (M1 and M3) and one of the heavy-load operating conditions (M2 but not M4). Increase in mutagenic activity was concentration dependent for all those 3 modes with the positive response at two highest concentrations (40 and 120 μg/plate). At the lowest studied concentration of 13.3 μg/plate, the mutagenic activity was found only for the engine fueled with B100 and operated at M2 condition (Figure 5).

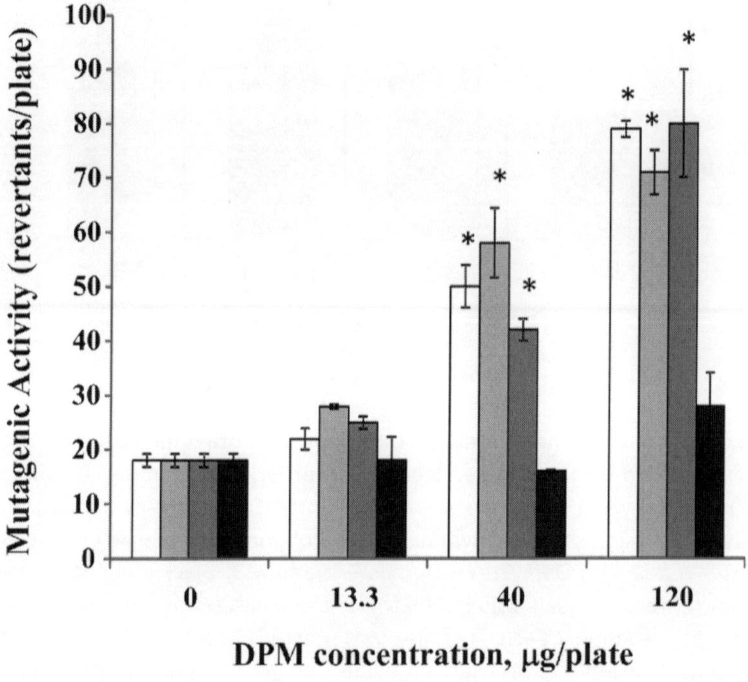

Figure 4. Effect of engine operating conditions on mutagenic activity of B50 samples collected for the engine fitted with muffler. B50 samples (0, 13.3, 40 or 120 μg/plate) were incubated at 37o C with the *Salmonella typhimurium* TA 98 without S9 microsomal activation. Clear columns – mode M1; light gray columns – mode M2; Dark gray columns – mode M3; Black columns – mode M4. Data represent mean values (+SEM) of the average number of revertant colonies per sample. Each sample was tested twice in two separate experiments. *positive responses, as evidenced by the number of revertant colonies being at least two-fold greater than the respective control value.

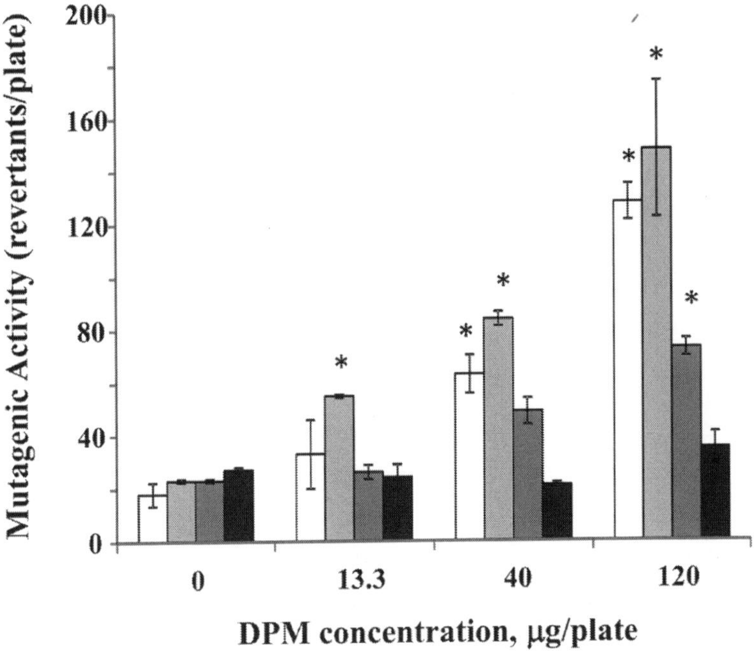

Figure 5. Effect of engine operating conditions on mutagenic activity of B100 samples collected for the engine fitted with muffler. B100 samples (0, 13.3, 40 or 120 µg/plate) were incubated at 37o C with the *Salmonella typhimurium* TA 98 without S9 microsomal activation. Clear columns – mode M1; light gray columns – mode M2; Dark gray columns – mode M3; Black columns – mode M4. Data represent mean values (+SEM) of the average number of revertant colonies per sample. Each sample was tested twice in two separate experiments. *positive responses, as evidenced by the number of revertant colonies being at least two-fold greater than the respective control value.

Effect of diesel oxidation catalyst (DOC) on mutagenic activity of DPM

The effects of DOC on mutagenic activity of DPM were studied using samples collected for all three fuels and the engine operated at one light-load (M1) and one heavy-load (M2) condition (Table 1, Figure 6). The strongest effects of DOC were observed when the engine was fueled with B100 and operated at mode M1 (Figure 6A). The results for two highest sample concentrations, 40 µg/plate and 120 µg/plate, showed that introduction of this specific DOC in the exhaust system resulted in 59% and 69% decrease in mutagenic activity, respectively. In the cases involving B100 and lowest sample concentrations (13.3 µg/plate), the samples collected for the engine equipped with DOC and operated at M2 exhibited 45.5% lower mutagenic activity than the corresponding sample obtained for the muffler-only configuration (Figure 6B). When the engine was fueled with B50, the effects of DOC were observed only for M2 conditions, demonstrating 12% and 16% decrease in mutagenic activity for two highest sample concentrations, respectively (Figure 6B). No significant effects of DOC were observed when the engine was fueled with ULSD.

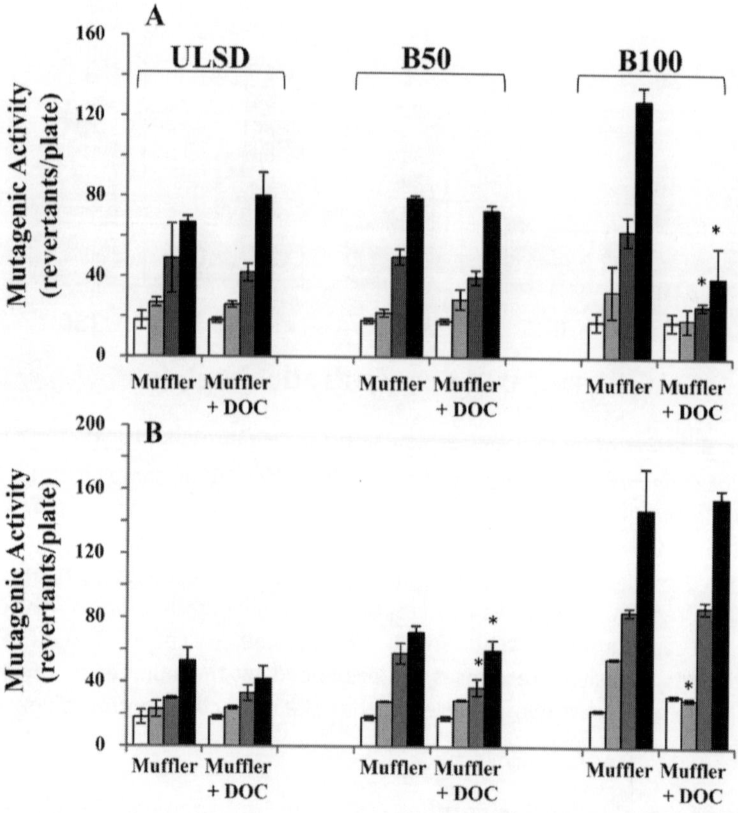

Figure 6. Effect of DOC on mutagenic activity of DPM samples collected for the engine operated at 2 steadystate modes M1 (A, light load) and M2 (B, heavy load). ULSD, B50 or B100 samples (0, 13.3, 40 or 120 μg/plate) were incubated at 37o C with the *Salmonella typhimurium* TA 98 without S9. Clear columns – control samples (PBS exposure); light gray columns – exposure with 13.3 μg/ plate of DPM; Dark gray columns – exposure with 40 μg/ plate of DPM; Black columns – exposure with 120 μg/ plate of DPM. Data represent mean values (+SEM) of the average number of revertant colonies per sample. Each sample was tested twice in two separate experiments. *p<0.05, vs the engine equipped with muffler only.

DISCUSSION

This study was conducted to investigate the mutagenic effects of soy-based FAME biodiesel. The effects of B50 and B100 were compared to those of ULSD. These effects were examined for four steady state engine operating conditions and two exhaust aftertreatment configurations using *Salmonella typhimurium* TA98 (frame-shift mutation) assay without metabolic activation (S9).

This is well-known that emissions are influenced by both the engine (and exhaust aftertreatment system) as well as the fuel being combusted [46]. Accordingly, results of our study clearly indicate that mutagenic activity of DPM was strongly affected by the fuel formulation, engine operating conditions, and exhaust aftreatment. For the engine fueled with ULSD, only the DPM samples collected for light-load engine operating conditions were mutagenic. These results are consistent with the previous studies [45,46] and can possibly be explained by the lower brake thermal efficiency of the engine, less completed combustion, and consequently elevated emission of organic compounds at the light-load in comparison with the high-load engine operating conditions. In contrast, samples collected for the engine fueled with B50 or B100, the mutagenic activity was observed for both light-load and heavy-load engine operating conditions. In our study, the increase in mutagenic activity correlated with the increased level of the fraction of organic carbon in these samples [23]. Samples with the higher fraction of organic carbon also exhibited the higher genotoxic activity for the strain TA98 without S9 microsomal activation. For example, DPM samples obtained for M1 and M2 operating conditions had respectively 64% and 168% higher organic carbon content when B100 was used in place of ULSD [23]. Consequently, the B100 samples exhibited 91% and 179% higher mutagenic activity than corresponding ULSD samples. The DPM generated at heavy-load engine operating condition (M4) that was characterized with the lowest fraction of organic carbon of all samples collected [23] showed no positive mutagenic activity. The mutagenic activity increased with fraction of biodiesel in the fuel for three modes (M1-M3). Similarly, the results produced by the other groups [30,31], indicate a strong dependency of the number of revertants on the engine speed. In our study, the strongest increases in mutagenicity were triggered by B100 at the high-speed modes M1 and M2. In agreement with other studies [17,25], carbon analysis of the primary filter showed positive correlation between fraction of organic carbon and fraction of biodiesel in the fuel [23].

It is well established that particulate matter produced by combustion of petroleum diesel and biodiesel are a complex mixture of different compounds [5,48]. Regardless of the reduction in the total mass of particulate matter, the soluble organic fraction of the emitted particles is commonly found at a higher level in the exhaust of diesel engines fueled with biodiesel. Many studies have found that soluble organic fraction (SOF) of DPM contains polynuclear aromatic hydrocarbons (PAH) and nitro- PAH that are known to be genotoxic and probably are strong contributors to induce gene mutation in bacteria [9,49-52]. Some of the PAH measured in the DPM, like benzo(a)pyrene, fluoranthene or benzo(ghi)perylene (chemically reactive, non-volatile), emitted at the greater levels in the biodiesel than diesel exhaust [53]. These PAHs are known to be mutagenic and highly carcinogenic. The positive results without metabolic activation confirm that components like substituted PAH contribute significantly to mutagenic effects of DPM. This effect of direct acting mutagens is attributed to substituted PAH, such as nitro-PAH [5,54-56]. Nitro-PAH could be formed by a reaction with the NOx present in diesel engine emissions. These substances display a strong direct mutagenicity while there parent compounds exhibit no or less mutagenicity [5,8,57,58]. The combustion of biodiesel in a diesel engine usually increases the release of nitrogen oxides [59]. An increase in NO_2 concentration in the B50 exhaust samples [23] corresponded with the higher mutagenicity observed for B50 samples as compared to ULSD. Furthermore, Biswas *et al.,* (2009) showed correlations between oxidative potential and organic carbon from DPM [60]. In our study we have an increased percentage of the organic carbon fraction for biodiesel samples in comparison with diesel exhaust. Higher concentrations of organic compounds and transition metals (Co, Cu, Ni, Zn) in the biodiesel exhaust than in diesel exhaust [61] might contribute to the higher oxidative potential of biodiesel DPM and thus to its higher *in vitro* mutagenicity. While the chemical characterization of the fuel exhaust is not part of the current study, future work focusing on deciphering the detailed compositions of the organic extracts of biodiesel and diesel exhaust and its effect on relation to the mutagenic activity, is highly prompted.

Benefits of using a DOC to control toxicity of DPM generates at light– and heavy-load engine operating conditions were investigated in our study. When the DOC was used, a strong correlation between engine operating conditions and observed changes in mutagenic activity of DPM was detected. DOC significantly decreased the mutagenic activity of DPM by 69% in the case of B100 and light-load (M1) engine operating conditions. A decrease in mutagenicity was also observed when engine equipped with DOC was fueled with B50 and heavily loaded. The decrease in mutagenicity can be explained by the fact that DOC effectively removed hydrocarbons and the organic carbon fraction of DPM [23]. Mutagenicity was slightly increased by the B100 exhaust (120 mg/plate) at high load mode M2 with a great increase in NO_2 concentration [23], which might be responsible for synthesizing even more genotoxic materials, nitro-PAH, under these conditions. Our results are similar to those from the different studies showing that changes depended mainly on engine load modes [34,62,63]. They suggested that high engine load increases NOx in the exhaust and NOx cannot effectively be reduced by a DOC, while high NOx levels lead to an increased formation of nPAH.

CONCLUSIONS

In conclusion, the neat and blended soy-based FAME biodiesel fuels demonstrated the potential to substantially reduce the mine air concentrations of elemental carbon (EC) and total DPM mass. However, the DPM generated by diesel engine fueled with neat or blended biodiesel was found to have higher mutagenic potency, as determined by Salmonella assays than DPM generated by the same engine fueled with ULSD. This increase in mutagenicity might be attributed to the higher presence of organic carbon in the biodiesel DPM. Since our results demonstrated that the evaluated DOC was effective in controlling OC level and mutagenic potential of DPM, we would recommend using optimized DOCs on the diesel engines particularly those fueled with biodiesel. Further research is needed to investigate the health effect of biodiesel exhaust as well as efficiency of the DOC and other exhaust aftertreatment devices in removal of specific potentially toxic species.

AUTHORS' CONTRIBUTIONS

ERK was interpreting the data and prepared the manuscript, XCS was completed the experiments and analyzed data, MJK was involved in experimental design and completing experiments, ABB was involved in experimental design of the study and collecting diesel/biodiesel exhaust particles, AAS was involved in coordinating the study, interpreting the data, and writing the manuscript.

ACKNOWLEDGEMENT

This work was supported by NIOSH 2927ZKCY.

REFERENCES

1. Balat, M. and Balat, H: **Progress in biodiesel processing.** *Applied Energy* 2010, **87**:1815-1835.
2. Krahl J., Seidel H., Bunger L: **Exhaust gas emissions of rapeseed oil based fuels and effects on environment and human health.** In: Chartier P. Ferrero GL, Henius UM, Hultberg S, Sachau J, Wiinblad M, editors. Proceedings of the 9th European bioenergy conference. Oxford: Elsevier; 1996, 1657-1661.
3. Graboski MS, McCormick RL: **Combustion of fat and vegetable oils derived fuels in diesel engines.** *Prog Energy Combust Sci* 1998, **24**: 125–164.
4. Bugarski, A.D., Janisko S., Cauda E.G., Noll, J.D., Mischler, S.E. **Controlling Exposure - Diesel Emissions in Underground Mines.** *Society for Mining, Metallurgy, and Exploration,* 2012, ISBN-13: 9780873353601.
5. Bunger J, Krahl J, Schroder O, Schmidt L and Westphal GA: **Potential hazards associated with combustion of bio-derived versus petroleum-derived diesel fuel.** *Crit Rev Toxicol* 2012, **42**:732-50.
6. Krahl, J., Knothe, G.H., Munack, A., Ruschel, Y., Schroder, O., et al: **Comparison of Exhaust Emissions and Their Mutagenicity from the Combustion of Biodiesel, Vegetable Oil, Gas-to-Liquid and Petrodiesel Fuels.** *Fuel* 2009, **88**:1064-1069.
7. Mermelstein R, Kiriazides DK, Butler M, McCoy EC and Rosenkranz HS: **The extraordinary mutagenicity of nitropyrenes in bacteria**. *Mutat Res* 1981, **89**:187-96.

8. Ohe T: **Mutagenicity of photochemical reaction products of polycyclic aromatic hydrocarbons with nitrite**. *Sci Total Environ* 1984, **39**:161-75.

9. Rivedal E, Myhre O, Sanner T and Eide I: **Supplemental role of the Ames mutation assay and gap junction intercellular communication in studies of possible carcinogenic compounds from diesel exhaust particles**. *Arch Toxicol* 2003, **77**:533-42.

10. Silverman DT, Samanic CM, Lubin JH, Blair AE, Stewart PA, Vermeulen R, Coble JB, Rothman N, Schleiff PL, Travis WD, Ziegler RG, Wacholder S and Attfield MD: **The Diesel Exhaust in Miners study: a nested case-control study of lung cancer and diesel exhaust**. *J Natl Cancer Inst* 2012, **104**:855-68.

11. Attfield MD, Schleiff PL, Lubin JH, Blair A, Stewart PA, Vermeulen R, Coble JB and Silverman DT: **The Diesel Exhaust in Miners study: a cohort mortality study with emphasis on lung cancer**. *J Natl Cancer Inst* 2012, **104**:869-83.

12. IARC**: diesel engine exhaust carcinogenic**. *Cent Eur J Public Health* 2012, **20**:120-138.

13. EPA **A comprehensive analysis of biodiesel impact on exhaust emissions.** *Draft technical report.*Washington, DC: U.S. Environmental Protection Agency, 2012. EPA420-P-02-001.

14. Schultz MJ, Fields KG, Atchison DJ, Gerbec EJ, Haney RA, et al. **Using bio-diesel fuels to reduce DPM concentrations; DPM results using various blends of bio-diesel fuel mixtures in a stone mine.** In: *Ganguli R, Bandopadhyay S, eds. Proceedings of the 10th U.S./North American Mine Ventilation Symposium.* 2004, Leiden, Netherlands: Balkema

15. Tomko DM, Stackpole RP, Findlay CD, Pomroy WH. **Metal/nonmetal diesel particulate matter rule.**In: Hardcastle S, McKinnon DL, **eds.** *Proceedings of the 13th U.S./North American Mine Ventilation Symposium.* 2010, Sudbury, ON, Canada: MIRARCO—Mining Innovation, 143–148.

16. Durbin, D.T., Cocker, D.R. III, Sawant, A.A., Johnson, K., Miller, J.W., et al: **Regulated emissions from biodiesel fuels from on/off-road applications.** *Atmos. Environ* 2007, **41**:5647-5658.

17. McDonald, J.F., Cantrell, B.K., Watts, W.F., and Bickel, K.L: **Evaluation of a soybean oil based diesel fuel in an underground gold mine.** *CIM Bulletin* 1997, **91**: 91-95.

18. McCormick, R.L. **Effects of biodiesel on NOX emissions. Presented at Air Resources Board Biodiesel Workshop.** *National Renewable Energy Laboratory*, Golden CO, 2005. [URL=http: //www.nrel.gov/ vehiclesandfuels/npbf/pdfs /38296.pdf

19. Watts, W.F., Spears, M., and Johnson, J. **Evaluation of biodiesel fuel and oxidation catalyst in an underground mine.** *Report to DEEP Technical Committee.* 1998.

20. Chen YC and Wu CH: **Emissions of submicron particles from a direct injection diesel engine by using biodiesel**. *J Environ Sci Health A Tox Hazard Subst Environ Eng* 2002, **37**:829-43.

21. Bugarski, A.D., Schnakenberg, G.H., Noll, J.D., Mischler, S.E., Patts, L.D., et al: **Effectiveness of selected diesel particulate matter control technologies for underground mining applications: Isolated zone study, 2003.** *Report of Investigations 9667.* U.S. Department of Health and Human Services, DHHS (NIOSH) Publication No. 2006-126, 2006.

22. Bugarski, A.D., Schnakenberg, G.H., Mischler, S.E., Noll, J.D., Patts L.D., et al: **Effectiveness of selected diesel particulate matter control technologies for underground mining applications: isolated zone study.** *Report of Investigations 9668.* U.S. Department of Health And Human Services, Public Health Service, Centers for Disease Control and Prevention, National Institute for Occupational Safety and Health, Pittsburgh Research Laboratory, Pittsburgh, PA, 2006.

23. Bugarski AD, Cauda EG, Janisko SJ, Hummer JA and Patts LD: **Aerosols emitted in underground mine air by diesel engine fueled with biodiesel**. *J Air Waste Manag Assoc* 2010, **60**:237-44.

24. Swanson KJ, Madden MC and Ghio AJ: **Biodiesel exhaust: the need for health effects research**.*Environ Health Perspect* 2007, **115**:496-9.

25. Purcell, D.J., McClure, B.T., McDonald, J., Basu, H.N: **Transient testing of soy methyl ester fuels in an indirect injection,**

compression ignition engine. *J. Am. Oil Chem. Soc* 1996, **73**:381-388.

26. Ames BN, Lee FD and Durston WE: **An improved bacterial test system for the detection and classification of mutagens and carcinogens.** *Proc Natl Acad Sci U S A* 1973, **70**:782-6.

27. Ames BN, McCann J and Yamasaki E: **Methods for detecting carcinogens and mutagens with the Salmonella/mammalian-microsome mutagenicity test.** *Mutat Res* 1975, **31**:347-64.

28. Carraro E, Locatelli AL, Ferrero C, Fea E and Gilli G: **Biological activity of particle exhaust emissions from light-duty diesel engines.** *J Environ Pathol Toxicol Oncol* 1997, **16**:101-9.

29. Bagley, S.T. and Gratz, D: **Evaluation of biodiesel fuel and a diesel oxidation catalyst in an underground metal mine, part 3 – biological and chemical characterization.** *Final Report. Submitted to the DEEP Technical Committee* 1998, **32**:1183–1191.

30. Bunger J, Krahl J, Franke HU, Munack A and Hallier E: **Mutagenic and cytotoxic effects of exhaust particulate matter of biodiesel compared to fossil diesel fuel.** *Mutat Res* 1998, **415**:13-23.

31. Bunger J, Krahl J, Baum K, Schroder O, Muller M, Westphal G, Ruhnau P, Schulz TG and Hallier E:**Cytotoxic and mutagenic effects, particle size and concentration analysis of diesel engine emissions using biodiesel and petrol diesel as fuel.** *Arch Toxicol* 2000, **74**:490-8.

32. Bunger J, Muller MM, Krahl J, Baum K, Weigel A, Hallier E and Schulz TG: **Mutagenicity of diesel exhaust particles from two fossil and two plant oil fuels.** *Mutagenesis* 2000, **15**:391-7.

33. Bünger J, Krahl J, Stein H, Schröder O, Müller M, et al: **Particle filters reduce emission and mutagenicity of diesel exhaust in regular operation but can cause an increase during the regeneration phase.** *Naunyn Schmiedeberg's Arch Pharmacol* 2004, **369**:145.

34. Bunger J, Krahl J, Weigel A, Schroder O, Bruning T, Muller M, Hallier E and Westphal G: **Influence of fuel properties, nitrogen oxides, and exhaust treatment by an oxidation catalytic**

converter on the mutagenicity of diesel engine emissions. *Arch Toxicol* 2006, **80**:540-6.

35. Bunger J, Krahl J, Munack A, Ruschel Y, Schroder O, Emmert B, Westphal G, Muller M, Hallier E and Bruning T: **Strong mutagenic effects of diesel engine emissions using vegetable oil as fuel**. *Arch Toxicol* 2007, **81**:599-603.

36. Kado NY, Huang TL, Kuzmicky PA. Truck-in-the-Park II: **Chemical and bioassay analyses of diesel and biodiesel particulate matter.** *Final Report,* EDG–95–7561. Montana Department of Environmental Quality. Helena, Montana, 2001.

37. Kado NY and Kuzmicky PA: **Bioassay analyses of particulate matter from a diesel bus engine using various biodiesel feedstock fuels**. *Final Report*. Report 3 in a series of 6. National Renewable Energy Laboratory U.S. Department of Energy, Office of Scientific and Technical Information, Oak Ridge, TN USA, 2003.

38. Krahl J, Munack A, Ruschel Y, Schröder O, Bünger J. **Exhaust gas emissions and mutagenic effects of diesel fuel, biodiesel and biodiesel blends.** *SAE Technical Paper* 2008, **2008**:01–2508.

39. Turrio-Baldassarri L, Battistelli CL, Conti L, Crebelli R, De Berardis B, Iamiceli AL, Gambino M and Iannaccone S: **Emission comparison of urban bus engine fueled with diesel oil and 'biodiesel' blend**. *Sci Total Environ* 2004, **327**:147-62.

40. Bugarski AD, Schnakenberg GH, Jr., Hummer IA, Cauda E, Janisko SI and Patts LD: **Effects of diesel exhaust aftertreatment devices on concentrations and size distribution of aerosols in underground mine air**. *Environ Sci Technol* 2009, **43**:6737-43.

41. MSHA (Mine Safety and Health Administration). *National diesel inventory*. 2009. http://lakegovprod2.msha.gov/DieselInventory/ViewDieselInventoryExternal.aspx. |

42. Maron DM and Ames BN: **Revised methods for the Salmonella mutagenicity test**. *Mutat Res* 1983, **113**:173-215.

43. Liu YQ, Keane M, Ensell M, Miller W, Kashon M, Ong TM, Mauderly J, Lawson D, Gautam M, Zielinska B, Whitney K, Eberhardt J and Wallace W: **In vitro genotoxicity of exhaust**

emissions of diesel and gasoline engine vehicles operated on a unified driving cycle. *J Environ Monit* 2005, **7**:60-6.

44. McMillian, M.H., Cui, M., Gautam, M., Keane, M., Ong, T., et al: **Diesel engine loading, speed, and fuel affect bacterial gene mutation activity induced by solvent extract of exhaust particulate material.** *SAE Technical Paper* 2002,2002-01-1699

45. Shi, X-C., Keane, M.J., Ong, T.-M., Harrison, J.C., Slaven, J.E., et al: **Diesel exhaust particulate material expression of in vitro genetoxic activities when dispersed into a phospholipids component of lung surfactant.** *Journal of Physics* 2009, Conference Series **151** 012021.

46. McClellan RO, Hesterberg TW and Wall JC: **Evaluation of carcinogenic hazard of diesel engine exhaust needs to consider revolutionary changes in diesel technology**. *Regul Toxicol Pharmacol*2012, **63**:225-58.

47. Lu T, Huang Z, Cheung CS and Ma J: **Size distribution of EC, OC and particle-phase PAHs emissions from a diesel engine fueled with three fuels**. *Sci Total Environ* 2012, **438**:33-41.

48. Kittelson DB: **Engines and nanoparticles: a review.** *J Aerosol Sci* 1998, **29**: 575–588.

49. Zielinska B, Sagebiel J, McDonald JD, Whitney K and Lawson DR: **Emission rates and comparative chemical composition from selected in-use diesel and gasoline-fueled vehicles**. *J Air Waste Manag Assoc* 2004, **54**:1138-50.

50. Moller M, Hagen I and Ramdahl T: **Mutagenicity of polycyclic aromatic compounds (PAC) identified in source emissions and ambient air**. *Mutat Res* 1985, **157**:149-56.

51. Pederson TC and Siak JS: **The role of nitroaromatic compounds in the direct-acting mutagenicity of diesel particle extracts**. *J Appl Toxicol* 1981, **1**:54-60.

52. Schuetzle D: **Sampling of vehicle emissions for chemical analysis and biological testing**. *Environ Health Perspect* 1983, **47**:65-80.

53. Kado NY, Okamoto RA, Kuzmicky PA: **Chemical and bioassay analyses of diesel and biodiesel particulate matter: pilot**

study. *Final report for The U.S. Department of Energy and The Renewable Energy Report Library,* 1996.

54. Hayakawa K, Nakamura A, Terai N, Kizu R and Ando K: **Nitroarene concentrations and direct-acting mutagenicity of diesel exhaust particulates fractionated by silica-gel column chromatography**. *Chem Pharm Bull (Tokyo)* 1997, **45**:1820-2.

55. Westerholm R, Christensen A, Tornqvist M, Ehrenberg L, Rannug U, Sjogren M, Rafter J, Soontjens C, Almen J and Gragg K: **Comparison of exhaust emissions from Swedish environmental classified diesel fuel (MK1) and European Program on Emissions, Fuels and Engine Technologies (EPEFE) reference fuel: a chemical and biological characterization, with viewpoints on cancer risk**. *Environ Sci Technol* 2001, **35**:1748-54.

56. Heeb NV, Schmid P, Kohler M, Gujer E, Zennegg M, Wenger D, Wichser A, Ulrich A, Gfeller U, Honegger P, Zeyer K, Emmenegger L, Petermann JL, Czerwinski J, Mosimann T, Kasper M and Mayer A: **Secondary effects of catalytic diesel particulate filters: conversion of PAHs versus formation of nitro-PAHs**. *Environ Sci Technol* 2008, **42**:3773-9.

57. HEI. Health Effects Institute: **Diesel exhaust: A critical analysis of emissions, exposure, and health effects.** *A special report of the institute's diesel working group.* USA: Cambridge; 1995.

58. Victorin K: **Review of the genotoxicity of nitrogen oxides**. *Mutat Res* 1994, **317**:43-55.

59. Mauderly JL: **Health issues concerning inhalation of petroleum diesel and biodiesel exhaust.** In:*Plant Oils as Fuels: Present State of Science and Future Developments* (Martini N, Schell J, eds). Berlin:Springer, 1997, 92–103.

60. Biswas S, Verma V, Schauer JJ, Cassee FR, Cho AK and Sioutas C: **Oxidative potential of semi-volatile and non volatile particulate matter (PM) from heavy-duty vehicles retrofitted with emission control technologies**. *Environ Sci Technol* 2009, **43**:3905-12.

61. Jalava PI, Tapanainen M, Kuuspalo K, Markkanen A, Hakulinen P, Happo MS, Pennanen AS, Ihalainen M, Yli-Pirila P, Makkonen U, Teinila K, Maki-Paakkanen J, Salonen RO, Jokiniemi J and

Hirvonen MR:**Toxicological effects of emission particles from fossil- and biodiesel-fueled diesel engine with and without DOC/POC catalytic converter**. *Inhal Toxicol* 2010, **22 Suppl 2**:48-58.

62. Tanaka Y., Hihara T., Nagata M., Azuma N., Ueno A: **Modeling of diesel oxidation catalyst.** *Ind Eng Chem Res* 2005, **44**:8205-8212.

63. Westphal GA, Krahl J, Munack A, Ruschel Y, Schroder O, Hallier E, Bruning T and Bunger J:**Mutagenicity of diesel engine exhaust is eliminated in the gas phase by an oxidation catalyst but only slightly reduced in the particle phase**. *Environ Sci Technol* 2012, **46**:6417-24.

Chapter 2

Design of Push–Pull System to Control Diesel Particular Matter Inside a Dead-End Entry

Yi Zheng[1], Magesh Thiruvengadam[1], Hai Lan[2], Jerry C. Tien[3]

[1]*Department of Mining & Nuclear Engineering, Missouri University of Science and Technology, Rolla, MO 65401, USA*

[2]*Clean Air Power Inc., Poway, CA 92064, USA*

[3]*Division of Mining and Resources Engineering, Department of Civil Engineering, Monash University, Clayton Campus, Wellington Road, Clayton, VIC 3800, Australia*

ABSTRACT

Diesel particulate matter (DPM) is considered to be carcinogenic after prolonged exposure. With more diesel-powered equipment used in underground mines, miners' exposure to DPM has become an increasing concern. This paper used computational fluid dynamics method to study the DPM dispersion in a dead-end entry with loading operation. The effects of different push–pull ventilation systems on DPM distribution were evaluated to improve the working conditions for underground miners. The four push–pull systems considered include: long push and short pull tubing; short push and long pull tubing, long push and curved pull tubing, and short push and curved pull tubing. A species transport model with buoyancy effect was used to examine the DPM dispersion pattern with unsteady state analysis. During the 200 s of loading operation, high DPM levels were identified in the face and dead-end entry regions. This study can be used for mining engineer as guidance to design and setup local ventilation, select DPM control strategies and for DPM annual training for underground miners.

Keywords: Diesel particulate matter; Computational fluid dynamics; Ventilation; Underground condition; Push–pull system

1 INTRODUCTION

For underground mines, self-propelled diesel equipment that does not require power cable or constant charging batteries is preferred because working faces usually cover extensive areas where these facilities are not available. However, emission from the tailpipe and its subsequent distribution in the underground mine are of growing concerns for miners.

Diesel particulate matter (DPM) is the particulate by-product of diesel exhaust and it can exist in different modes with different size distributions (5 nm to 10 μm). Due to its very small size and very complex adsorb ability (more than 1800 different organic compounds and potentially toxic hydrocarbons were identified (CFR 2001)), it can be breathed into the alveolar region of the lungs of miners and cause acute health problems such as asthma, eye and nose irritation, headaches and nausea (Kahn and Orris 1988; Wade and Newman 1993; Rundell et al. 1996) to long term carcinogenic effects (NIOSH 1988; EPA 2002).

For underground coal mines, diesel engines used underground are divided into three categories under Mine Safety and Health Administration (MSHA) regulations: "permissible", "nonpermissible heavy-duty equipment, generators, and compressors" and "nonpermissible light-duty equipment". Equipment under each category is required to emit no more than a certain amount of DPM per hour (30 CFR 72.D 2014); otherwise, it will not be allowed to operate underground.

For underground metal/non-metal mines, MSHA regulations limit a miner's personal exposure to DPM no more than 160 μg/m^3 of total carbon (TC) for an average eight-hour equivalent full shift (effective from May 20, 2008) (30 CFR 57.5060 2014). Till today, there are still mines that cannot meet this regulation limit.

To control DPM hazards, two types of strategies have been commonly used. One is DPM reduction and removal before it is released from the engine tailpipe, which includes proper diesel engine selection and maintenance (McGinn 1999; Anyon2008; McGinn et al. 2010), use of alternative fuels (Zannis et al. 2008; Bugarski et al. 2010), and exhaust gas treatment devices (Shah et al. 2007; Bugarski et al.2009), e.g., diesel particulate filters (DPF). The other is through control measures after DPM is discharged into the environment—mine ventilation system, an enclosed equipment cab with filtered breathing air (environmental cab), personal protective equipment, and administrative controls (Cecala et al. 2005; Noll et al.2008; MSHA 2013).

Experience (Bugarski et al. 2011) showed that no single strategy can solve all DPM problems and a combination of several measures needs to be implemented in the field to attain compliance. To achieve an effective, efficient, and economical control scheme, an understanding of DPM behaviour in mining environment can be very useful in selecting the control strategies and training the miners. Numerical simulations using CFD can be used for that purpose by visualizing DPM distribution based on laboratory experiments and field studies.

CFD simulations have been successfully used in mining research to detect spontaneous combustion and apply inertisation in gob areas (Yuan and Smith 2007; Ren et al. 2005), study airflow patterns and gas concentrations in continuous miner operations or heading development (Sullivan and Heerden 1993; Hargreaves and Lowndes 2007; Wala et al. 2007; Kollipara et al. 2012; Torno et al. 2013), investigate scrubber intake designs for longwall dust control (Ren and Balusu 2008), and estimate a mine's damage status by tracer gas and simulation after a disaster (Xu et al. 2013). Simulation of DPM dispersion in underground mines was carried out by Zheng and Tien (2008), in which DPM was considered to behave like a gas. Subsequent study showed that it gave good quantitative agreement with practical accuracy for the DPM distribution and successfully identified the DPM affected areas above the threshold limit (Zheng et al. 2011). In the present study, DPM emission was also treated as a gas to examine its diffusion inside an underground single dead end entry.

Push–pull systems use both blower fan/tubing and exhaust fan/tubing systems, which can combine the effects of both single systems to lower the DPM levels in the dead-end face area. However, if not properly designed, the combination cannot provide the maximum outcomes and may interfere with each other. In this study, the effect of four different push–pull ventilation systems on DPM distribution inside a single dead-end entry was studied for a loading operation. CFD method was used to perform the evaluation and an optimum result was reached based on the four scenarios designed and assumptions made. This study can be used for mining engineer as guidance to design and setup of local ventilation. The high DPM regions revealed by the simulation can also be used for selection of DPM control strategies and DPM annual training for underground miners.

2 PROBLEM STATEMENT AND CFD MODELLING

2.1 Statement of the problem

A schematic of the computational domain for each push–pull ventilation design is shown in Figs. 1, 2, 3, 4. The four design cases considered in this study were: case (1), long push and short pull tubing (Fig. 1); case (2), short push and long pull tubing (Fig. 2); case (3), long push and curved pull tubing (Fig. 3); and case (4), short push and curved pull tubing (Fig. 4). The main entry measured 6 m in width, 5 m in height and 131 m in length, while the dead-end measured 6 m in width, 5 m in height, and 90 m in length. The three stub entries were evenly developed inside the dead-end entry and had the same cross-sectional dimensions as the main entry with 15 m depth. The main entry had of 19.35 m^3/s (41000 cfm) of fresh air flowing from the left to the right. The blower fan at the inlet of the push tubing was set to provide 8.02 m^3/s (17000 cfm) of fresh air into the face area. The exhaust fan at the outlet of the pull tubing drew the diesel exhaust mixture at a rate of 9.44 m^3/s (20000 cfm) from the face area and released it into the main entry.

For case (1), the push tubing extends into the dead-end entry for approximately 77 m while the pull tubing extended for approximate 67 m. For case (2), the push tubing extended for 70 m into the dead-end and the pull tube extended for approximate 81 m. In case (3), the length of the push tubing remained the same as in case (1) but the pull tubing, in addition to extending for 78 m into the dead-end, curved for an additional 12 m into the face area. In case (4), the length of the push tubing remained the same as in case (2) but the pull tubing, in addition to extending for 78 m into the dead-end, curved for an additional 12 m into the face area. The diameter of both the push and pull tubing was 0.8 m for all the cases and were placed 4.3 m above the ground.

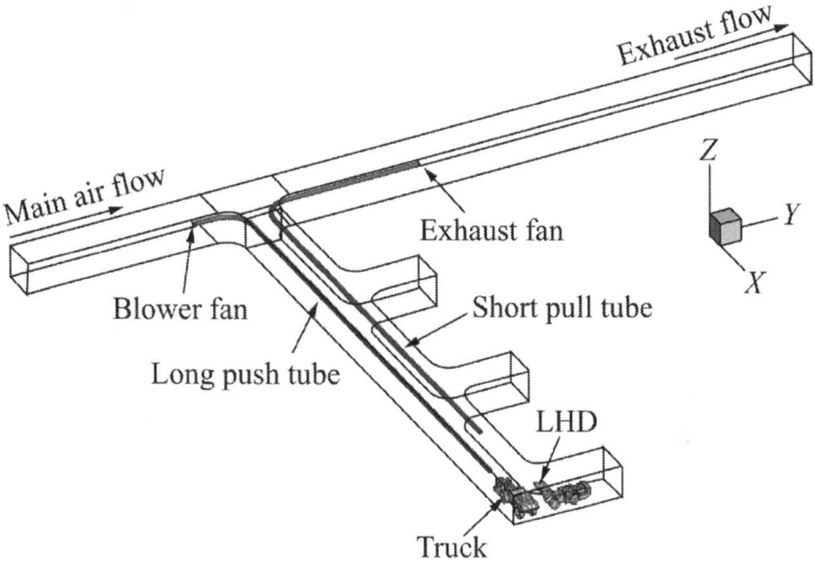

Figure1. Computational domain with long push and short pull ventilation system

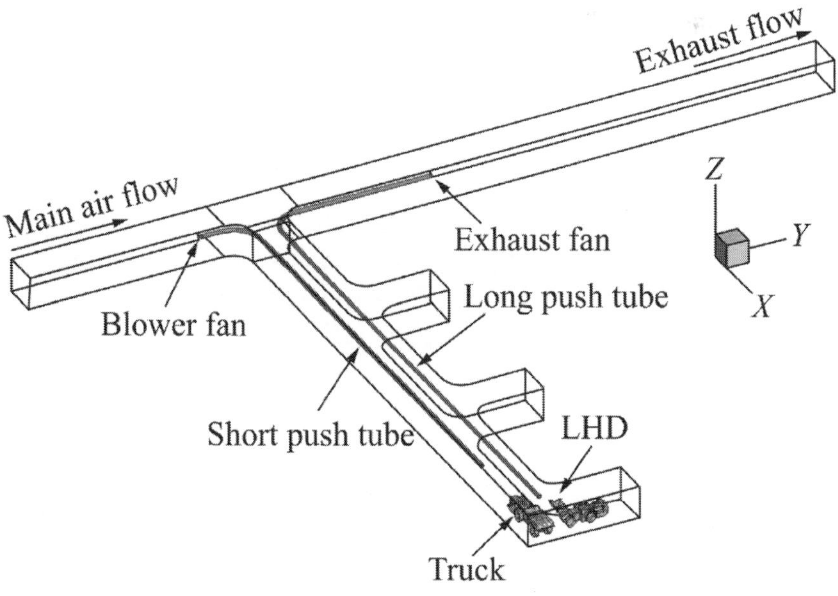

Figure 2. Computational domain with short push and long pull ventilation system

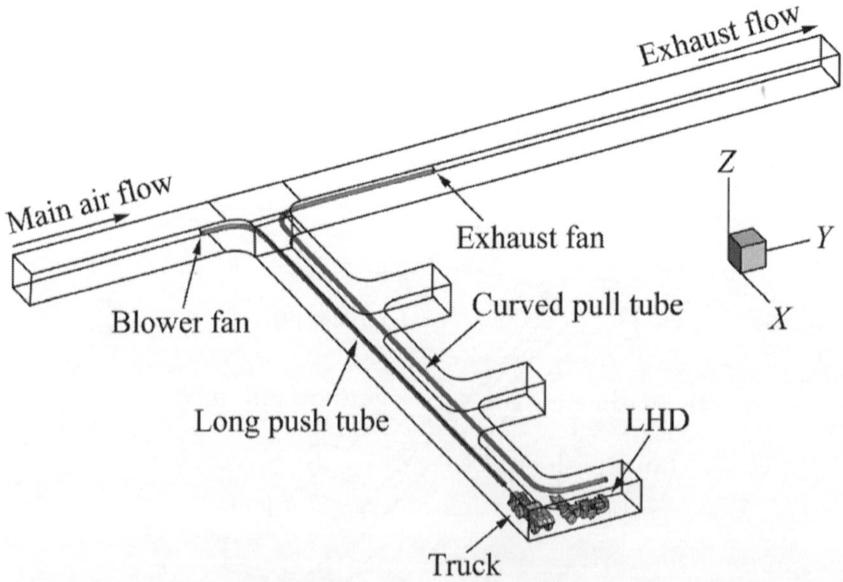

Figure 3. Computational domain with long push and curved pull ventilation system

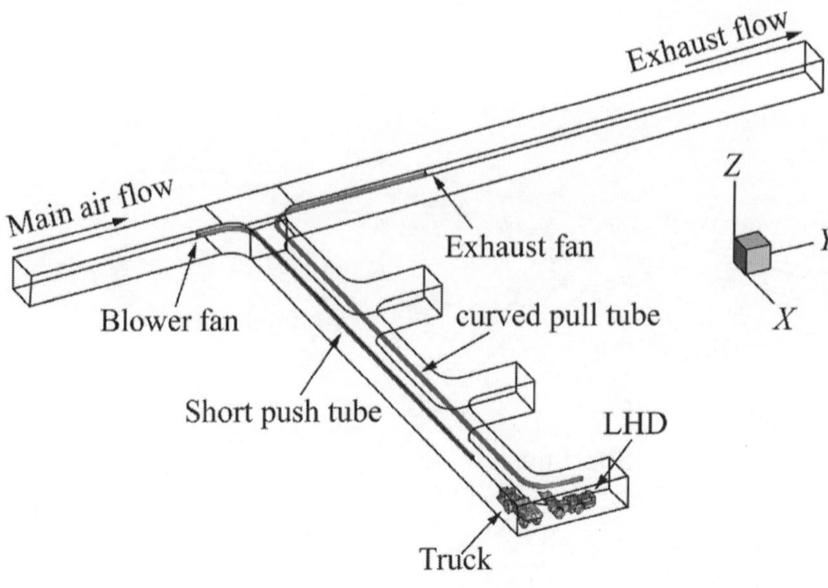

Figure 4. Computational domain with short push and curved pull ventilation system

2.2 Assumptions

Based on Zheng's studies (Zheng and Tien 2008; Zheng et al. 2011), this CFD simulation was conducted under the following assumptions: (1) DPM was treated as gas and the material selected as the representative for DPM was n-octane vapor (C_8H_{18}). The chemical reactions between species are not considered in this study; (2) both air and DPM are incompressible; (3) the flow in the domain is fully turbulent; (4) Both the loader and truck are stationary.

2.3 CFD modelling

Three-dimensional incompressible unsteady turbulent continuity, momentum, and energy equations, along with standard k-ε turbulent and non-reacting transport Eqs. (2 species, DPM and air) were solved using ANSYS FLUENT CFD software. The species transport model, available in FLUENT, was used to determine the DPM distribution pattern.

Due to the multiple cases covered in this section, all of the boundary conditions are summarized in Table 1. For the LHD, the tailpipe is located at the right rear of the LHD and pointing backward; for the truck, the tailpipe is placed at the right front of the vehicle, pointing to the floor. The parameters for the main ventilation and diesel vehicles are derived from Zheng's industrial field study (McGinn et al. 2004; Zheng et al. 2010) with only the low emission diesel engines. The detailed meanings of the boundary conditions are presented in other sections and in the FLUENT manual (ANSYS 2014).

Table 1. Summary of boundary conditions

	Boundary	Detailed settings
Main ventilation	Inlet	0.65 m/s, normal to boundary; DPM: 0 ppm
	Exit	Pressure outlet (0 Pa)
Diesel equipments	LHD tailpipe	24.1 m/s, normal to boundary; 594 K; DPM, 1.73 ppm
	Truck tailpipe	27.5 m/s, normal to boundary; 644 K; DPM, 2.0 ppm
Walls		No slip, adiabatic walls
Auxiliary ventilation	Push-tube	Inlet: fan ($\Delta p = 481$ Pa); outlet: interior
	Pull-tube	Inlet: interior; outlet: fan ($\Delta p = 800$ Pa)

In order to achieve accuracy in the simulation results, finer meshes were generated for the area close to diesel engines where high gradients existed. For all the models, about 1.5 million computational elements (cells) were generated. The unsteady flow calculations were made by using time step ($\Delta t = 0.1$ s) for the time period of 200 s (3 min and 20 s) for the loading operation.

3 RESULTS AND DISCUSSION

3.1 Long push and short pull tubing system (Case 1)

It can be seen from Fig. 5 that, when the pull tubing was shorter than the push tubing, the fresh air from the push tubing impinged on the rear wall of the dead-end and made a 90° turn to enter the face area. However, only a small portion of that fresh air reached the interior of the face area to ventilate the LHD emission. The remaining portion created a recirculation region in the rear section of the truck and in the frontal portion of the LHD, and then gradually migrated toward the main entry and inlet of the pull tubing. The distant location of the pull tubing made it difficult to remove all the exhaust mixture from the face area in a timely manner.

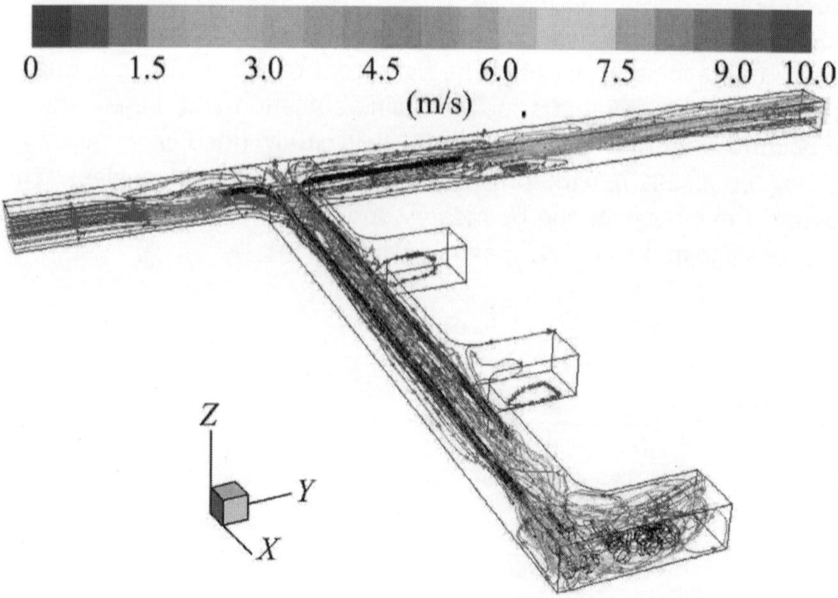

Figure 5. Pathlines colored by velocity magnitude demonstrating general flow features for long push short pull

The DPM distribution with long push tubing and short pull tubing is shown in Fig. 6. The push tubing was longer than the pull tubing by 10 m. The colored contours represent the diesel exhaust, with the DPM level above the regulation limit of 160 μg/m^3. The blower fan at the inlet of the push tubing provided 8.02 m^3/s (17000 cfm) of fresh air to the face area. This low-temperature fresh air mixed and cooled the high-temperature diesel exhausts of the LHD and truck engines and formed a DPM-air mixture. The exhaust fan at the outlet of the pull tubing sucked this exhaust mixture at a rate of 9.44 m^3/s (20000 cfm) and released it into the main entry. The high concentration DPM completely engulfed the active face area in the dead-end, except for small regions behind the truck and in front of the LHD. Most of the remaining areas of the dead-end were also filled with diesel fumes near the roof region by the end of the loading operation due to the buoyancy effect. The operators of the LHD and the truck should use enclosed cabs to protect themselves from the harmful effects of DPM. This design of long push and short pull tubing system failed to ventilate the active face area effectively.

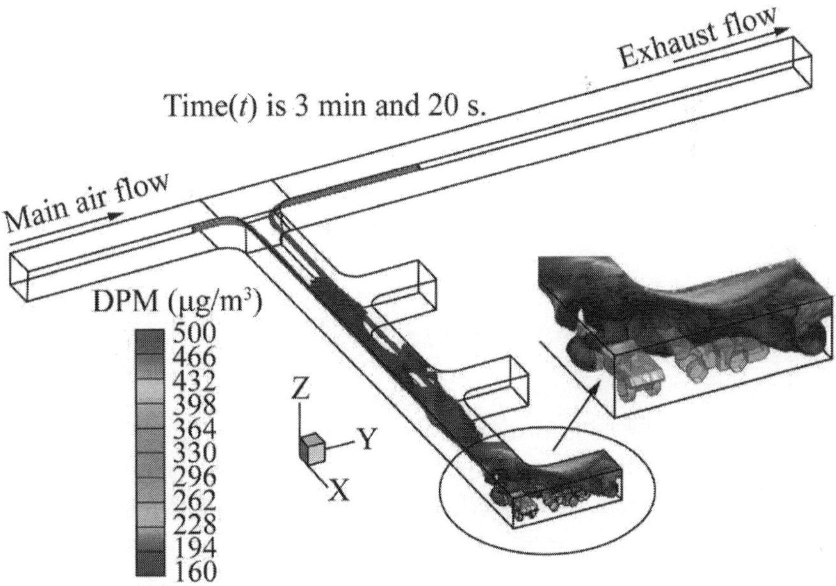

Figure 6. DPM distribution inside the single dead-end entry for long push short pull

3.2 Short push and long pull tubing system (Case 2)

In this case, the pull tubing was made longer than the push tubing, as shown in Fig. 7. With this modified design, the fresh air flow impinging on the rear wall of the dead-end created a strong vortex-like flow in the presence of pull tubing, as shown in the figure. Again, not enough fresh air reached the interior of the face area. But the closer location of the pull tubing to the face region helped to suck in more tailpipe emission of the LHD.

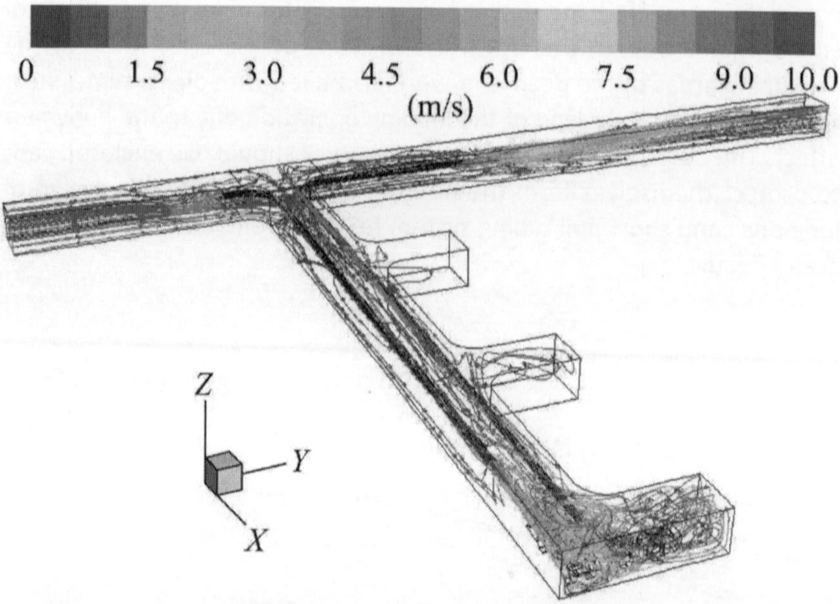

Figure 7. Pathlines colored by velocity magnitude demonstrating general flow features for short push long pull

Figure 8 shows the DPM distribution in the same dead-end with a short push and long pull tubing system for the LHD-truck loading operation. This modified design of short push and long pull tubing system effectively ventilated the face area, when compared with the long push and short pull tubing ventilation system. The DPM occupied a small region behind the tailpipe of the LHD, a small region around the tailpipe of the truck and the roof region near the face area due to the buoyancy force, as shown by the colored region. The miners working in this colored region should use personal protection instruments. The remaining areas of the dead-end were free of any DPM above the regulatory limit.

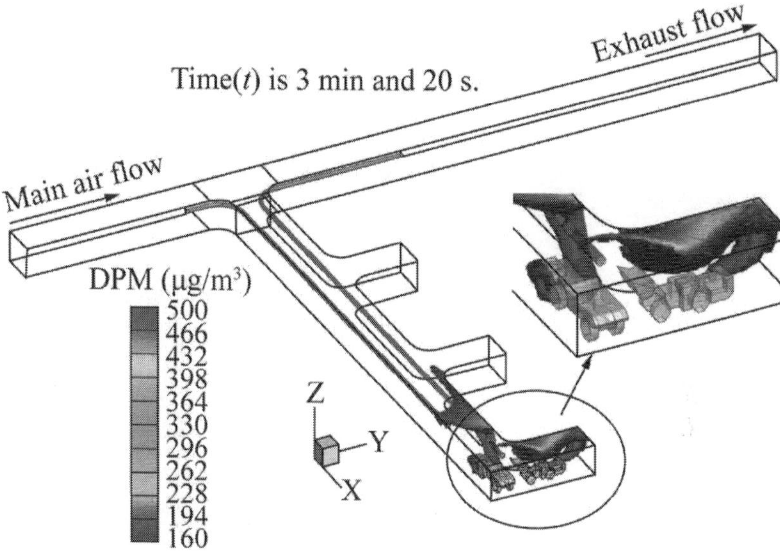

Figure 8. DPM distributions inside the single dead-end entry for short push long pull

3.3 Long push and curved pull tubing system (Case 3)

It can be seen in Fig. 9 that two recirculation regions were created in the face area: one is around the truck; the other is encircling the LHD. When compared with Fig. 5, there was a large improvement in the ventilation of the face area at the back of LHD around the inlet region of the pull tubing.

This system is similar to case 1 except that the pull tubing was extended and made to curve inside the face area. The resultant DPM distribution in the dead-end is shown in Fig. 10. When compared with case 1, there was a dramatic improvement in the DPM distribution in the dead-end. There was no high concentration DPM accumulation in the dead-end other than in the face area. However, this system did not perform better than the system in case 2 since there was significant DPM accumulation inside the face area where the LHD was located. It seems that a part of the DPM plume from LHD was missing the inlet of the pull tubing due to the high velocity of air from the push tubing. Miners in the colored region should use personal protection instruments during their working.

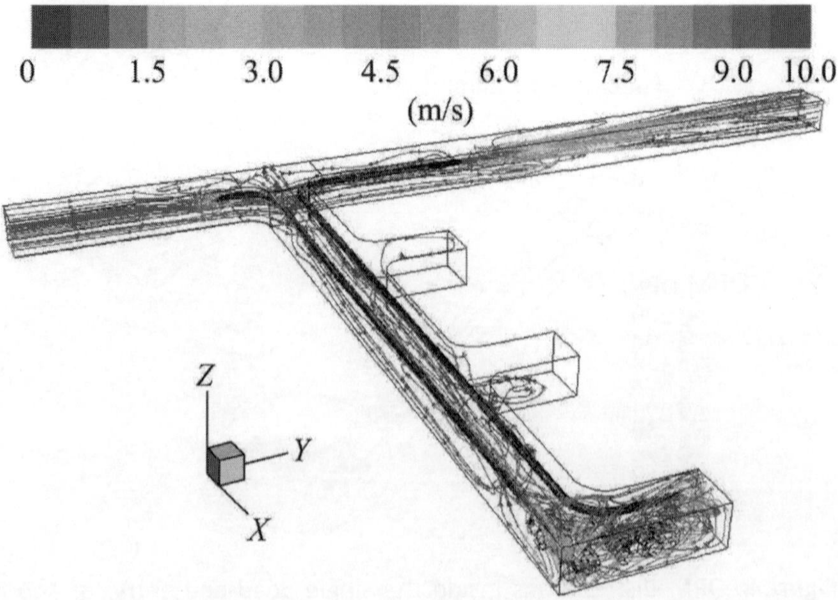

Figure 9. Pathlines colored by velocity magnitude demonstrating general flow features for long push curved pull

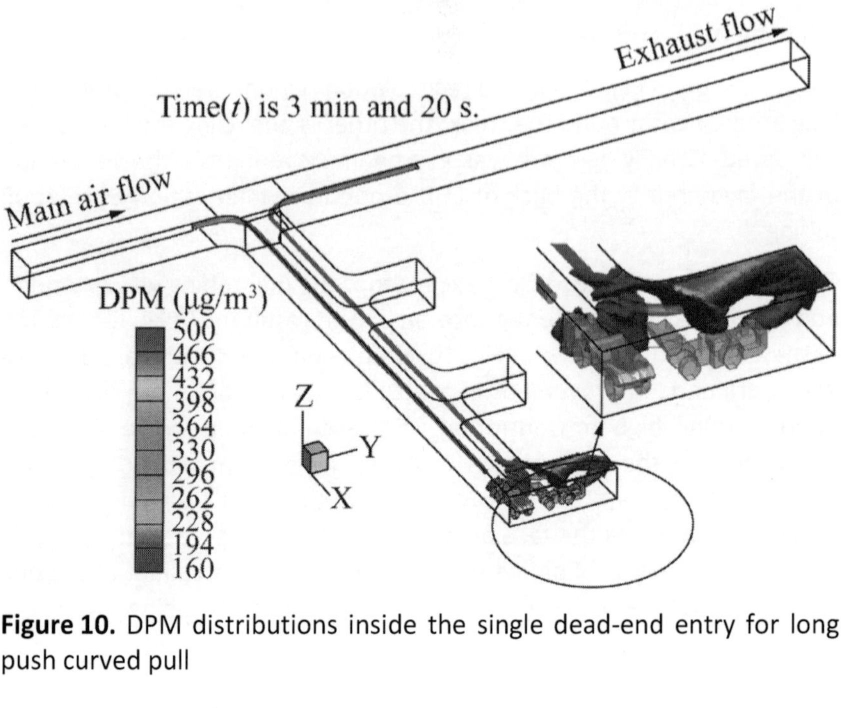

Figure 10. DPM distributions inside the single dead-end entry for long push curved pull

3.4 Short push and curved pull tubing system (Case 4)

As shown in Fig. 11, there was a single large recirculation region which effectively ventilated the entire face area.

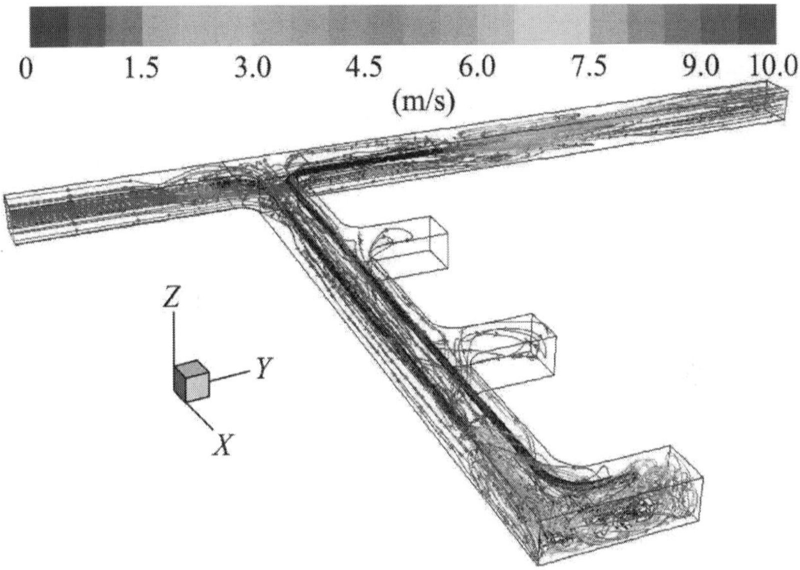

Figure 11. Pathlines colored by velocity magnitude demonstrating general flow features for short push curved pull

The DPM distribution, when a short push and long curved pull tubing auxiliary ventilation system was used for the LHD-truck loading operation, is shown in Fig. 12. This design is similar to case 2 with the short push and long pull tubing design except that the long pull tubing was made to curve for an additional 12 m into the working face area of the dead-end. A comparison of the design of the short push and long curve pull tubing system with short push and long straight pull tubing system showed only a negligible difference for the DPM affected areas near the tailpipe of the LHD and the truck. However, it significantly reduced the DPM accumulation in the roof region of the face area. As mentioned above, miners in the colored regions should use personal protection instruments during their working.

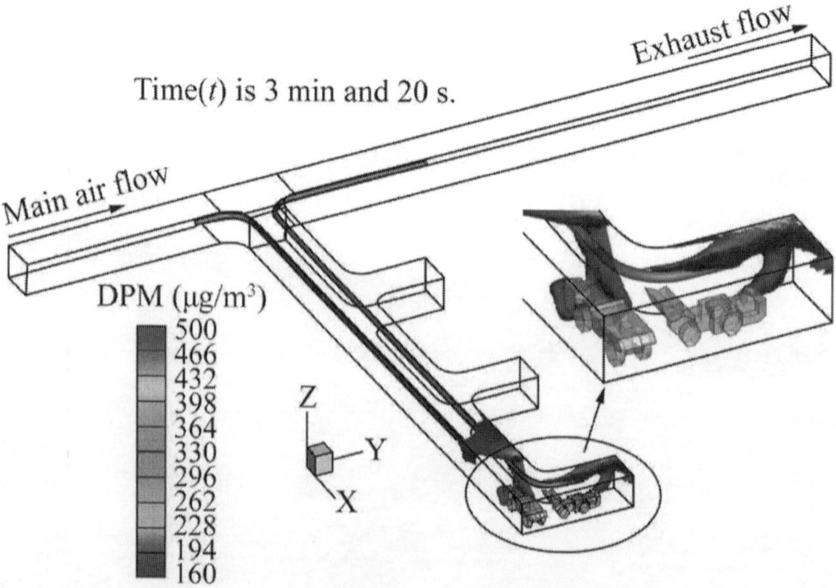

Figure 12. DPM distributions inside the single dead-end entry

4 COMPARISON OF DIFFERENT PUSH–PULL TUBING DESIGNS

A comparison of the different designs of the push–pull ventilation systems was made by plotting the area weighted DPM values at cut cross-sectional planes inside the dead-end. The cross-sectional planes are shown in Fig. 13 for the long push and short pull tubing ventilation system. Similar cross-sectional planes were created for other ventilation systems and are not shown here due to space limitations. The performance of each push–pull design system was evaluated based on its DPM dilution capability in the face area.

The overall performance evaluation of different push–pull designs was made by plotting the area-weighted average, the maximum, and minimum values of DPM at these cross-sectional planes against the distance from the face, as shown in Figs. 14, 15, and 16. The distance of the cross-sectional planes from the interior face area was evaluated, as shown in Fig. 13. It can be seen from Fig. 14, that case 4, the short push and curved pull tubing system, performed the best with the minimum average DPM value inside the face area, while case 1, the long push and short pull tubing system, performed the worst with the maximum average values inside the face area, when compared with other push–pull deisgns. Although case 3, the long push and curved pull system, resulted in minimum average values in the remaining areas of the dead-end, when compared with other systems, in the important face region where miners were working, case 4, the short push and curved pull tubing design, performed the best.

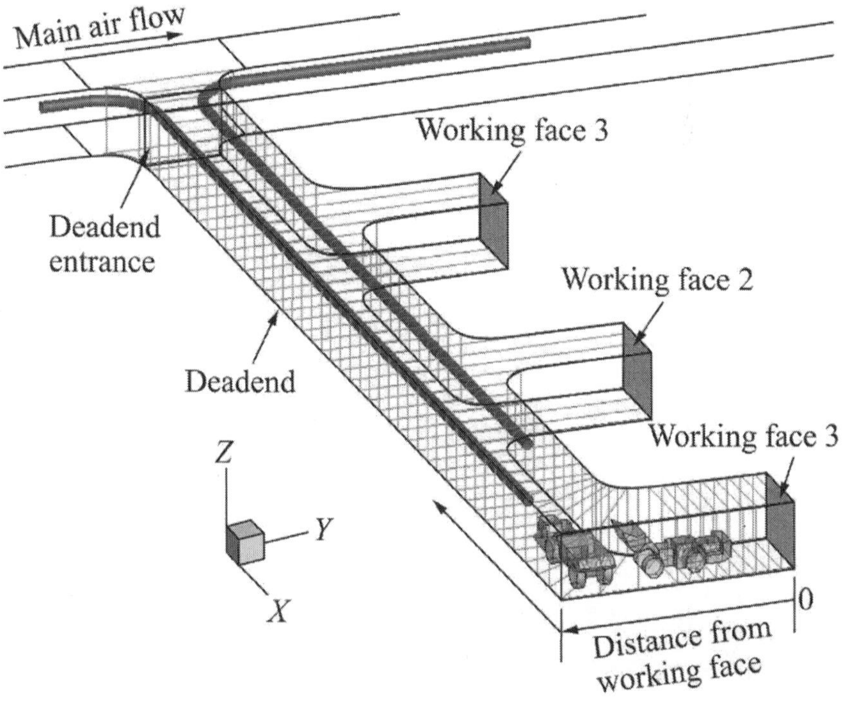

Figure 13. Schematic of the cross-sectional planes inside the dead-end

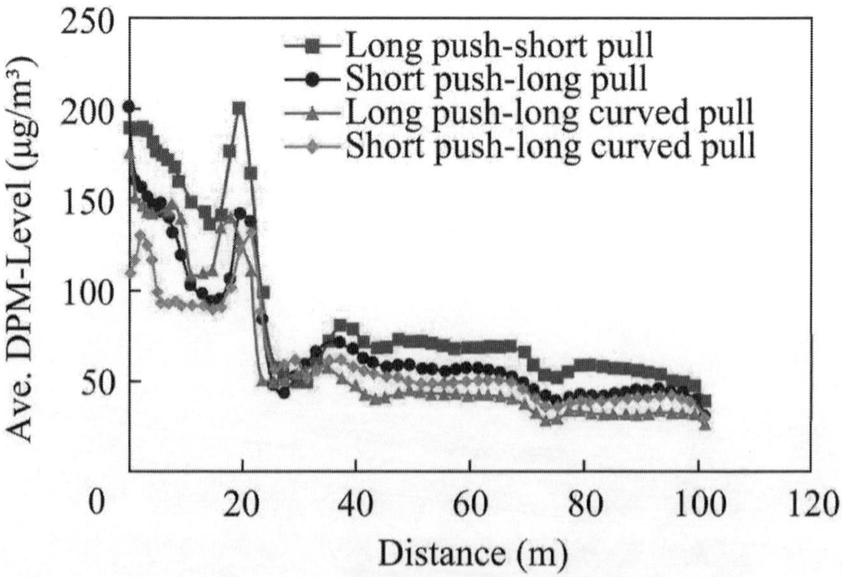

Figure 14. Comparison of average DPM values away from the face

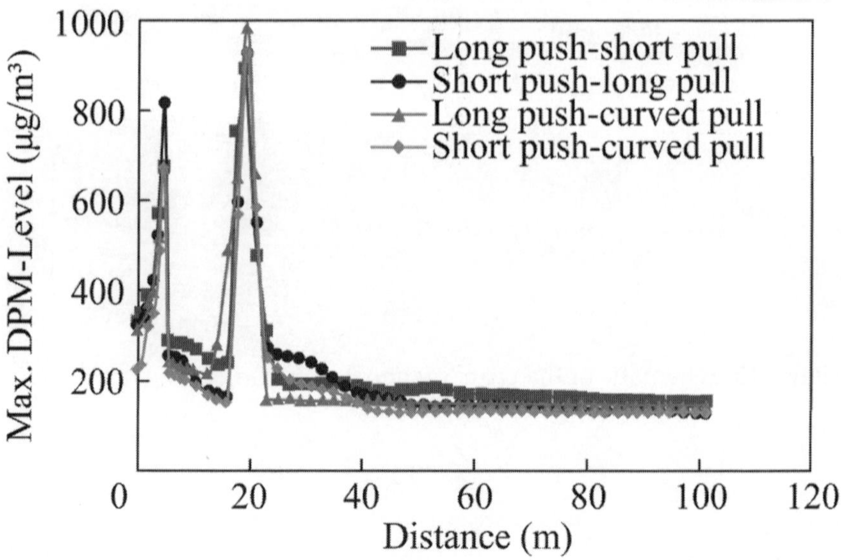

Figure 15. Comparison of the maximum DPM values away from the face

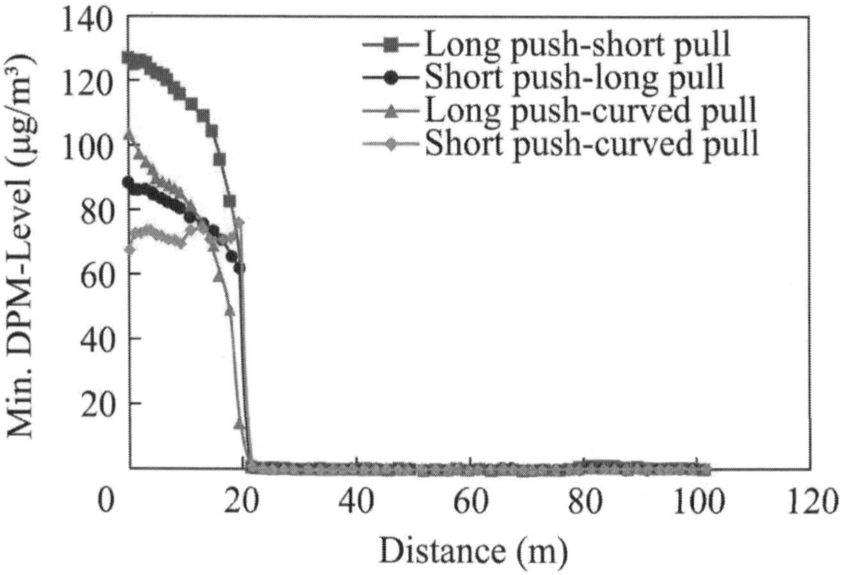

Figure 16. Comparison of the minimum DPM values away from the face

This fact was further substantiated with the plots of the maximum and minimum value of DPM at the cross-sectional planes, as shown in Figs. 15 and 16. Although the plot of the maximum value (Fig. 15) showed negligible differences in the distribution, the plot of the minimum value (Fig. 16) showed that minimum DPM values were obtained in the face area in case 4, and the short push and curved pull tubing system, was used.

5 CONCLUSIONS

It was found out by this study that, although clean engines were used in the face area, the ventilation condition can play an essential role in controlling DPM under the regulation requirements. Without proper auxiliary ventilation, the face area and dead-end entry will eventually filled with high DPM fumes.

From the comparison study of these four push–pull ventilation systems, it was concluded that the long push tubing provided powerful airflow in the face region and the DPM air mixture occupied more face region due to the high mixing force. Short push tubing produced better forward airflow momentum to confine the DPM plume in the face region and not over mixing the DPM with airflow. On the contrary, long or curved pull tubing served better DPM removal capacity as it approached high DPM face region and sucked the contaminant air directly out of the working area. Short pull tubing cannot remove DPM effectively as its influence region was too far away from the high DPM regions.

Based on this study, it was concluded that the short push and curved pull tubing system was the best design of all four designs, and effectively ventilated the face area during the truck loading operation. In case of unpractical, the straight short push and long pull tubing system can also have close ventilation result.

ACKNOWLEDGMENTS

The authors wish to express their sincere gratitude to the financial support provided by the Western US Mining Safety and Health Training & Translation Center by the National Institute for Occupational Safety and Health (NIOSH) (Grant Number 1 R25 OH008319).

REFERENCES

1. 30 CFR 57.5060 (2014) Limit on exposure to diesel particulate matter. Code of Federal Regulations, Title 30, part 57.5060, 370.
2. 30 CFR 72.D (2014) Diesel particulate matter—underground areas of underground coal mines. Code of Federal Regulations, Title 30, part 72, subpart D, 489–491.
3. ANSYS (2014) FLUENT user's guide. Release 15. ANSYS, Inc., Canonsburg
4. Anyon P (2008) Managing diesel particle emissions through engine maintenance—an Australian perspective. In: Proceedings of the 12th U.S./North American mine ventilation symposium, Reno, 521–526
5. Bugarski AD, Schnakenberg GH Jr, Hummer JA (2009) Effects of diesel exhaust aftertreatment devices on concentrations and size distribution of aerosols in underground mine air. Environ Sci Technol 43(17):6737–6743

6. Bugarski AD, Cauda EG, Janisko SJ (2010) Aerosols emitted in underground mine air by diesel engine fueled with biodiesel. J Air Waste Manag Assoc 60(2):237–244

7. Bugarski AD, Janisko SJ, Cauda EG (2011) Diesel aerosols and gases in underground mines: guide to exposure assessment and control. Department of Health and Human Services, Centers for Disease Control and Prevention, National Institute for Occupational Safety and Health, Office of Mine Safety and Health Research, Pittsburgh and Spokane

8. Cecala AB, Organiscak JA, Zimmer JA (2005) Reducing enclosed cab drill operator's respirable dust exposure with effective filtration and pressurization techniques. J Occup Environ Hyg 2(1):54–63

9. CFR (2001) Code of federal regulation. US Government Printing Office, Office the Federal Register.

10. EPA (2002) Health assessment document for diesel engine exhaust. U.S. Environmental Protection Agency.http://cfpub.epa.gov/ncea/cfm/recordisplay.cfm?deid=29060

11. Hargreaves DM, Lowndes IS (2007) The computational modeling of the ventilation flows within a rapid development drivage. Tunn Undergr Sp Technol 22(2):150–160

12. Kahn G, Orris P (1988) Acute overexposure to diesel exhaust: report of 13 cases. Am J Ind Med 13(3):405–406

13. Kollipara VK, Chugh YP, Relangi DD (2012) A CFD analysis of airflow patterns in face area for continuous miner making a right turn cut. 2012 SME Annual Meeting & Exhibit (preprint), 626–633

14. McGinn S (1999) Maintenance guidelines and best practices for diesel engines, final report, Diesel emissions evaluation program (DEEP).

15. McGinn S, Grenier M, Gangal M, Rubeli B, Bugarski A, Schnakenberg G, Johnson R, Petrie D, Crowther G, Penney J (2004) Brunswick mine diesel particulate filter field study. Diesel emissions evaluation program (DEEP) final report of investigation.

16. McGinn S, Ellington R, Penney J (2010) Diesel emissions: mechanics' maintenance manual, Diesel emissions evaluation program (DEEP). http://www.camiro.org/DEEP/Technology%20Sum%20And%20Reports/mechanicsman.pdf

17. MSHA (2013) Practical ways to reduce exposure to diesel exhaust in mining—a toolbox, U.S. Department of Labor, Mine Safety and Health Administration.

18. NIOSH (1988) Carcinogenic effects of exposure to diesel exhaust. National Institute for Occupational Safety and Health (NIOSH), Department of Health and Human Services. http://www.cdc.gov/niosh/docs/88-116

19. Noll JD, Patts L, Grau R (2008) The effects of ventilation controls and environmental cabs on diesel particulate matter concentrations in some limestone mine. In: Proceedings of the 12th US/North American mine ventilation symposium, Reno, 463–468

20. Ren TX, Balusu R (2008) Innovative CFD modeling to improve dust control in longwalls. Coal 2008: Coal Operators' Conference, Wollongong, 137–142

21. Ren TX, Balusu R, Humphries P (2005) Development of innovative goad inertisation practices to improve coal mine safety. COAL 2005: Coal Operators' Conference, Brisbane, 315–322

22. Rundell B, Ledin MC, Hammarstrom U, Stjernberg N, Lundback B, Sandstrom T (1996) Effects on symptoms and lung function in humans experimentally exposed to diesel exhaust. Occup Environ Med 53(10):658–662

23. Shah SD, Cocker DR III, Johnson KC (2007) Reduction of particulate matter emissions from diesel backup generators equipped with four different aftertreatment devices. Environ Sci Techol 41(14):5070–5076

24. Sullivan P, Heerden JV (1993) The simulation of environmental conditions in continuous miner developments using computational fluid dynamics. J Mine Vent Soc S Afr 1993:2–11

25. Torno S, Toraño J, Ulecia M (2013) Conventional and numerical models of blasting gas behavior in auxiliary ventilation of mining headings. Tunn Undergr Sp Technol 34:73–81

26. Wade JF III, Newman LS (1993) Diesel asthma: reactive airways disease following overexposure to locomotive exhaust. J Occup Med 35(2):149–154

27. Wala AM, Vytla S, Huang G (2007) Mine face ventilation: a comparison against benchmark experiments for the CFD code validation. Miner Eng 59(10):49–55

28. Xu G, Luxbacher KD, Ragab S (2013) Development of a remote analysis method for underground ventilation systems using tracer gas and CFD in a simplified laboratory apparatus. Tunn Undergr Sp Technol 33:1–11

29. Yuan L, Smith AC (2007) Computational fluid dynamics modeling of spontaneous heating in longwall gob areas. Trans Soc Min Metall Explor 322:37–44

30. Zannis TC, Hountalas DT, Papagiannakis RG (2008) Effects of fuel chemical structure and properties on diesel engine performance and pollutant emissions: review of the results of four European research programs. SAESP 2185:1–36

31. Zheng Y, Tien JC (2008) DPM dispersion study using CFD for underground metal/nonmetal mines. In: Proceedings of the 12th U.S./North America Mine ventilation Symposium, Reno, 487–493

32. Zheng Y, Thiruvengadam M, Lan H, Tien JC (2010) Simulation of DPM distribution in a long single entry. In: 13th Mine Ventilation Symposium, Sudbury, 149–156

33. Zheng Y, Lan H, Thiruvengadam M, Tien JC (2011) DPM dissipation at MST's experimental mine and comparison with simulation. J Coal Sci Eng (China) 17(3):285–289

Chapter 3

A Short History of the Toxicology of Inhaled Particles

Ken Donaldson[1] and Anthony Seaton[2]

[1]MRC/University of Edinburgh Centre for Inflammation Research, ELEGI Colt

Laboratory, Queen's Medical Research Institute, 47 Little France Crescent

[2]Environmental and Occupational Medicine, University of Aberdeen, Aberdeen, UK.

ABSTRACT

Particle toxicology arose in order to understand the mechanisms of adverse effects of 3 major particle types that had historically exerted the greatest toll of ill-health—quartz, coal and asbestos. By the middle of the last century rat inhalation studies had been carried out and the pathology documented, but true mechanistic particle toxicology did not really take off until the 1970s when cell culture techniques became available. By the 1980s glass fibres were a major focus of interest and attempts to develop a structure-toxicity paradigm centred on biopersistence. In the 1990s environmental particles dominated the particle toxicology agenda and the cardiovascular system emerged as a target for inhaled particles, raising new challenges for particle toxicologists. We are currently in the era of nanotoxicology where a large and diverse range of new nanoparticles types are under scrutiny.

The aims of particle toxicology are generally those of conventional chemical toxicology, that is to investigate the characteristics of particles that determine their ability to cause harm and the pathobiological mechanism of that harm. Achievement of these aims should enable identification of the best metric to measure particles in order to manage their risk and to intervene in cases where exposure has occurred in order to try to prevent progression. Particle toxicology has problems that other branches of toxicology don't have, central to which is the variation amongst the particles that are being tested—a problem that continues to confound efforts to make generalisations about their toxicology. This has most recently been found in the sub-specialty of nanotoxicology where the idea that all nanoparticles are equal in toxicity has been slow to counter [1]. Many new recruits to particle toxicology have taken up this area of toxicology because of the flow of funding into nanotoxicology, and we feel it may be helpful to them to know something of the previous history of particle toxicology, of which they may not be aware. While there are other ways in which particles might have adverse effects, our definition of particle toxicology is restricted to inhalation toxicology, meaning the accidental inhalation of particles in workplaces and the environment. A short history of inhalation particle toxicology is therefore our aim, and complements shorter accounts by other authors [2–5]. This is a personal history involving our own perception of the landmarks; a different close observer of the last 40 years of particle toxicology might write a different history. In such a short history it was impossible to cover the work of all the particle toxicologists who have advanced our discipline and we apologise for any such omissions. A lengthier history would be necessary to correct these.

THE FIRST DESCRIPTIONS OF DUST-RELATED OCCUPATIONAL LUNG DISEASE

The first mentions of a link between exposure to dust particles and lung disease are to be found in the 15th and 16th centuries. Within a few years of the invention of the printing press, Georgius Agricola wrote about the theory and practice of metal mining and refining in 'On the nature of minerals' (De re metallica) (Figure 1).

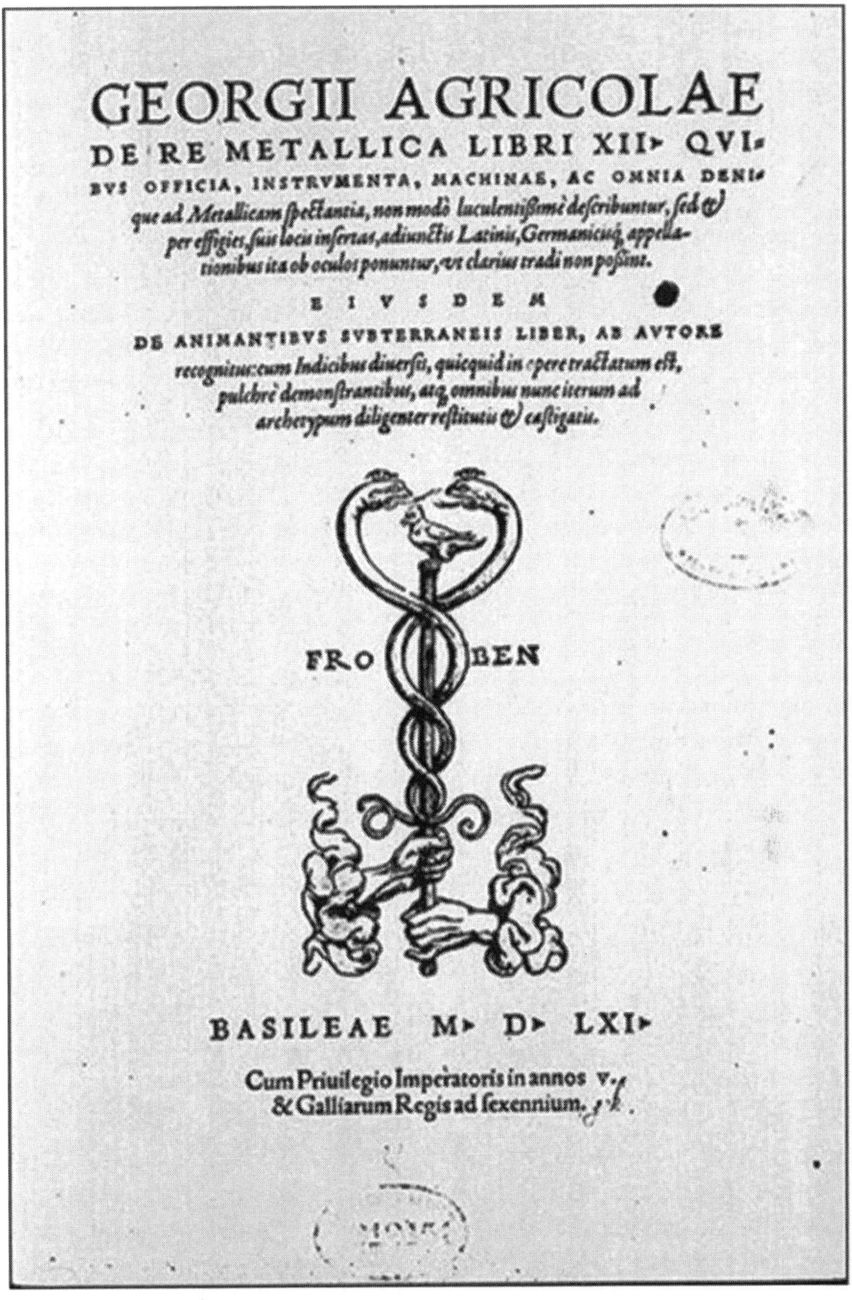

GEORGII AGRICOLAE

DE RE METALLICA LIBRI XII. QVIBVS OFFICIA, INSTRVMENTA, MACHINAE, AC OMNIA DENIque ad Metallicam spectantia, non modò luculentissimè describuntur, sed & per effigies, suis locis insertas, adiunctis Latinis, Germanicisq; appellationibus ita ob oculos ponuntur, ut clarius tradi non possint.

EIVSDEM

DE ANIMANTIBVS SVBTERRANEIS LIBER, AB AVTORE recognitus: cum Indicibus diuersis, quicquid in opere tractatum est, pulchrè demonstrantibus, atq; omnibus nunc iterum ad archetypum diligenter restitutis & castigatis.

FRO BEN

BASILEAE M. D. LXI.

Cum Priuilegio Imperatoris in annos V. & Galliarum Regis ad sexennium.

Figure 1. The front page of De re metallica (copyright expired; obtained through (Wikipedia commons).

Agricola described the theory and practice of mining including finding seams and working them. However he also described how air becomes stagnant in the mines and he described machines to blow fresh air into the mines, early ventilation systems. Along with accidents adversely affecting the miners he also mentioned how '*dust which is stirred by digging, penetrates into the windpipes and lungs and produces difficulty in breathing and the disease which the Greeks call asthma*'. It is important to point out that this is not our current use of the term asthma but a generic term for breathlessness. He stated that '*In the mines of the Carpathian mountains women are found who have married seven husbands, all of whom this terrible consumption has carried to a premature death*'. At that time no distinction was made between different lung diseases but later it was discovered that the miners were at risk from three diseases, silicosis, tuberculosis and lung cancer, all of which are likely to have contributed to their early demise [2]. Agricola did not link the dust exposure directly to the disease, although he recommends a primitive dust mask in the form of 'veils' that miners should wear. He is also regarded as the father of occupational hygiene in his description of methods of ventilating mines. Perhaps we should not be too surprised of the lack of simple observational correlation between dust in the air and lung disease since Agricola also wrote an entire book about the demons (De Animantibus Subterraneis) that inhabit mines. These ran about in the darkness chattering and turning out the miners' lamps but could be vanquished by fasting and prayer!

Paracelsus, best known for his observations on dose in toxicology and for his promotion of the first real cure, mercury, for syphilis, published the first book devoted to occupational disease in 1567. The term contrarian might have been coined for Paracelsus. In an extended 'gap year' following graduation in medicine he wandered around Europe meeting all sorts of people from gypsies to midwives and executioners and formed his own brand of folk medicine. On returning in 1526 he became town physician in Basel and lecturer in the University. Paracelsus soon started publicly burning the previous generation's medical textbooks and giving long, hugely popular lectures recounting his personal experiences; the success of these made him unpopular with his colleagues. Of relevance, he used his experience of smelting plants and mines that he visited during his travels and his work for mines in later years, to develop a theory of pathology in miners' lungs. He recognised and described the symptoms of miners' lung disease—

dyspnoea, cough and wasting. Like Agricola he failed to make the link to their exposures and instead evoked a theory involving '*...the powers of the stars, in that their peculiar characters are boiled out which settle on the lungs in different ways*' [2]. Concepts of medicine were dominated in those days by belief in disease being inflicted by demons or a consequence of an imbalance of the humours, so these explanations would not have been particularly strange to Paracelsus' contemporaries.

The person recognised as the father of occupational medicine was however Bernardino Ramazzini, who in 1700 wrote his text 'Diseases of workers'. His most important insight was to recognise that illness could have an environmental cause. There was a view that disease was sometimes caused by bad air (eg malaria from Italian, *mala aria*) but he extended this to a much wider range of factors. He also was one of the first doctors to concern himself with illness in poor working people. Amongst many insights on occupational diseases, for that time, he discussed ergonomics / bad posture, recommended that there should be good ventilation in dusty trades, spacious rooms with a good draught, and that workers should quit work at the first sign of any lung disease [3].

Despite these early insights, the working conditions of the poor, which included miners and those in dusty trades, were of no interest to legislators until the Industrial Revolution in the early 19th century and little was done to protect workers from danger until the early 20th century. From then on, in the developed world, steps have been taken to reduce exposures in workplaces, by which time exposures to the big three hazards, quartz (crystalline silica), coal and asbestos had exerted a huge toll in terms of particle-related lung disease, and this continues in industrially developing countries to the present day. Sometimes the scientific establishment took the risks rather too lightly. For example J.S Haldane who was honorary Director of the Mining Research laboratory of the University of Birmingham declared in 1923 that '*the inhalation of coal dust causes no danger to life but on the contrary gives protection against the development of tuberculosis*'. At that time silica was well known to cause lung disease but the respiratory illness among coal miners was thought to be a consequence of tuberculosis and quartz. Although he had a medical degree, Haldane, a physiologist, was most interested in acute problems such as gassing, explosions and pressure effects, and it was not until pathologists in Cardiff such as Cumming and Gough investigated the lungs of miners in the period 1930–50 that it became apparent that coal dust was toxic in its own right [6, 7]. This of course was also partly a consequence of the introduction of medical radiology in that era.

THE 'BIG THREE' OF PARTICLE TOXICOLOGY

Three particle types have been responsible for the greatest amount of lung disease over time and these are dealt with below.

Quartz or crystalline silica is one of the most common minerals in the earth's crust so that throughout the ages whenever mankind (and they were mostly, but not exclusively men) mined, quarried or worked the surface layers of the earth, they were exposed to quartz dust as the large crystals fractured into respirable particles. This toll of death from silicosis was recognised first in the 18th and early 19th centuries among knife grinders in Sheffield and men who made millstones, but undoubtedly went unnoticed from the time that men shaped arrow heads and axes from flint. Lung cancer in miners, subsequently shown to be due to radon exposure, was first described at the end of the 19th century. In the 1920s the Hawk's Nest tunnel incident in Gauley Bridge, West Virginia, produced a horrific toll of death and disease (Figure 2). Of the estimated 2,500 workers who worked in the tunnel, 764 died from acute silicosis and an additional 1,500 ultimately developed the disease [8]. Although the employers attempted to cover this up, the details can be read about owing to the fascinating detective work of Martin Cherniack in his book 'The Hawk's Nest Incident' [9]. Unfortunately, silicosis continues to take its toll of human lives, even in developed countries, though to a much greater extent in poorer countries with inadequate regulation of exposed trades.

Asbestos was first mined in substantial amounts in the late 1800 s in Canada and was soon being imported in millions of tons per annum from the major producers, Canada, Russia, South Africa and Australia into most developed countries, where it was incorporated into a huge number of products such as cement, tiles, fillers, brake linings, pipes and insulation (Figure 3). Like quartz, asbestos is crystalline and the crystals shatter along fracture planes to release thin fibres that are easily breathed in to the lungs. Asbestos causes a number of lung diseases including asbestosis, a fatal type of interstitial lung fibrosis, and lung cancer but also affects the pleura for reasons that are becoming better understood [10]. In the pleura it gives rise to pleural fibrosis, plaques and mesothelioma, the most feared of the asbestos diseases.

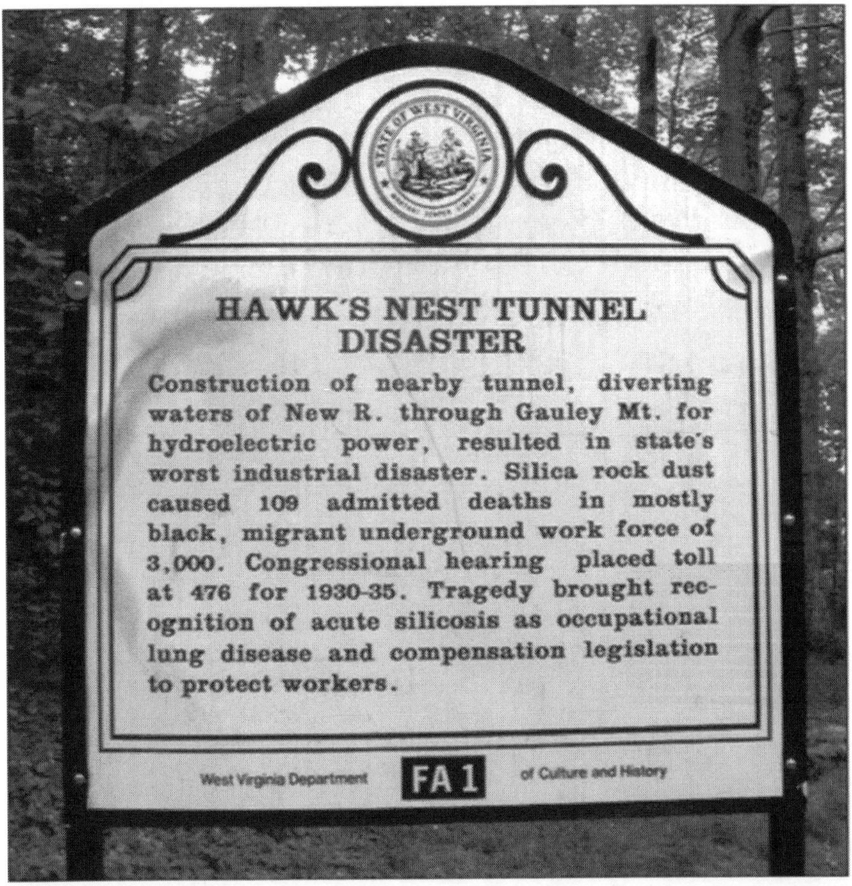

Figure 2. Hawk's Nest Tunnel memorial in Ansted, Fayette County, West Virginia, USA. (From the Historical Markers Database http://www.hmdb.org/marker.asp?marker=34417).

Pleural effects in humans, especially mesothelioma, have only been seen with asbestos and one other long fibre-shaped particle, erionite, that is not part of the asbestos family; pleural effects are not seen with other particles except in very advanced silicosis and mesothelioma is only seen with fibres. Mesothelioma is a slow-growing but highly malignant tumour that causes death usually within a year or two of diagnosis, with no known cure. Although asbestos is no longer used in most developed countries, in many countries it still is, and exposure still occurs from its removal or disturbance. In Western countries thousands of mesothelioma deaths occur every year owing to the use of asbestos until the 1980s and this is likely to continue for decades in most countries, particularly where amphibole asbestos has been used [11].

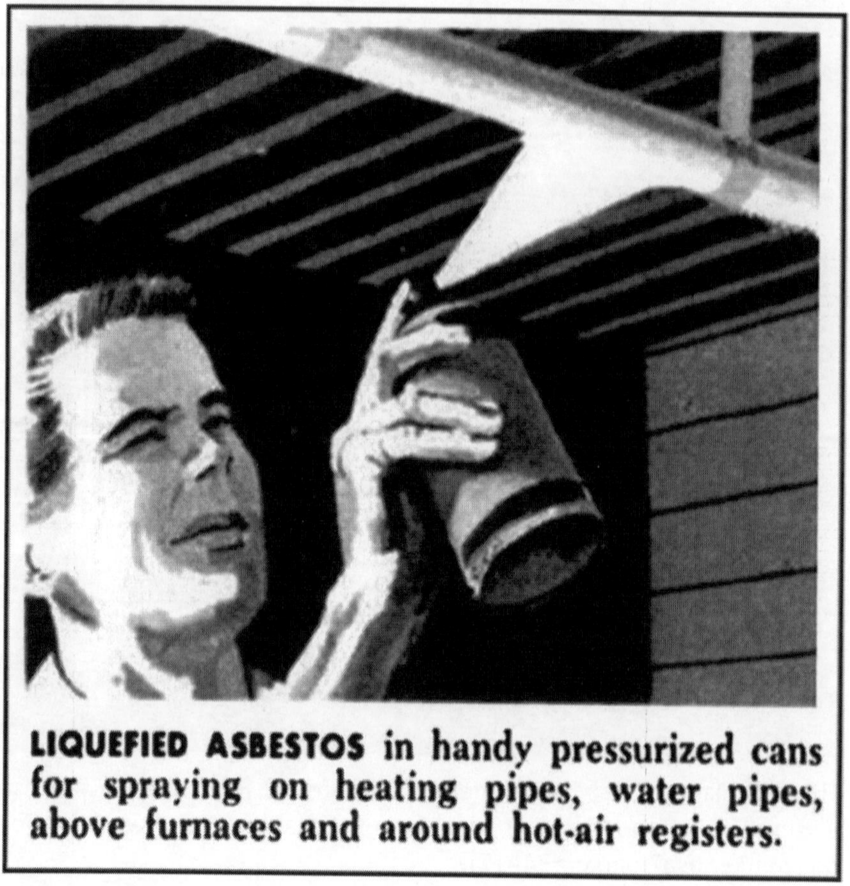

Figure 3. Advert for 'liquefied asbestos' showing ill-advised use of an asbestos spray by a man with no personal protective equipment. **(From Asbestos focus website** http://www.asbestosfocus.co.uk/apps/photos/photo?photoid=112479766).

Coal causes a form of pneumoconiosis (coal-worker's pneumoconiosis) that is usually less severe than silicosis. Coalmine dust comprises mostly low toxicity particles containing a variable mixture of carbon, quartz and silicates, and coalworker's pneumoconiosis is a lung disease caused by large accumulations of coal dust-laden macrophages with a modest local tissue reaction of mild fibrosis usually accompanied by emphysema (Figure 4). In a significant proportion of individuals it develops into progressive massive fibrosis, probably related to accumulation of very high lung burdens caused by the dust-induced inflammatory reaction in draining lymph nodes blocking clearance [12].

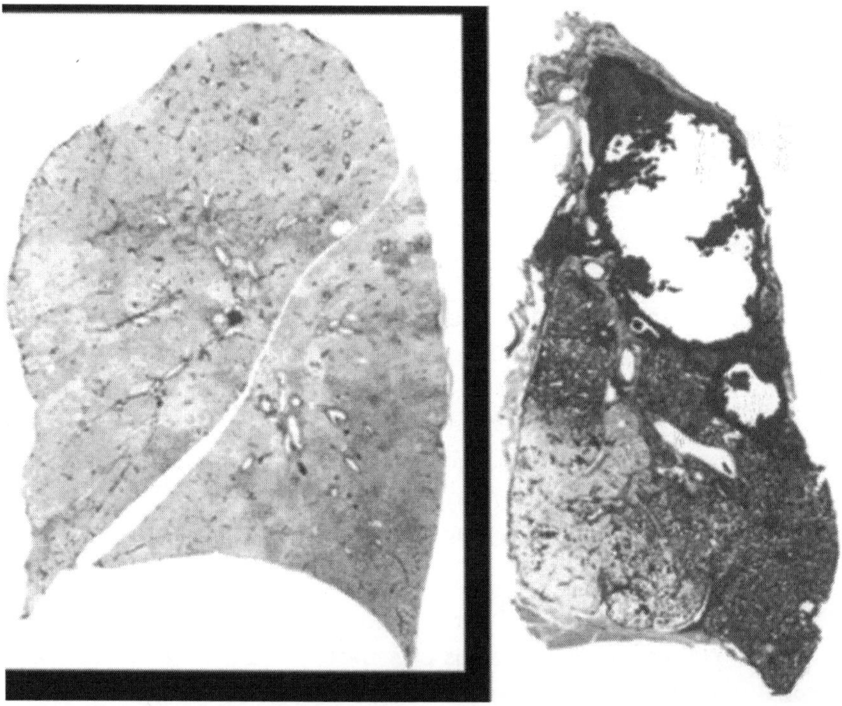

Figure 4. Gough Wentworth paper- mounted thick sections of the entire lung of a non-smoking non-miner (left) and a coal miner with cavitating progressive massive fibrosis (right). The left hand lung shows some centri-acinar dust accumulations in the upper parts related to air pollution but is otherwise normal. The miner's lung shows the dark pigmentation of coalmine dust with emphysema and almost total destruction of the upper lobe with apical pleural thickening. (with permission from Ingelheim portfolio 11 Lung sections).

In the middle of the 20th century the UK National Coal Board's 'Pneumoconiosis Field Research' developed an integrated approach to understanding the link between coalmine dust exposures and coal worker's pneumoconiosis, using epidemiologists, hygienists, pathologists and radiologists. Radiographs of coal workers were used to monitor prevalence of the disease and the risk of this was related to measured dust exposures of the miners. In addition, the extent of the disease in many of the miners' lungs was assessed at autopsy. The ambitious aim was to determine the elements of the coalmine dust (quartz, clays, coal itself) that were responsible for disease. This analysis was carried out in the Institute of Occupational Medicine in Edinburgh. This huge undertaking initially involved the development of a sampler (MRE114A) that was critical for determining exposures to respirable dust. In fact when the huge amounts of data were brought together, including more than 30,000 chest radiographs alone, the conclusion was clear—no single component of the coalmine dust mix was responsible and the total respirable dust was the metric that was best related to disease development. The exposure-response relationship calculated by Michael Jacobsen and colleagues led to agreement between trades unions, employers and government on a coalmine dust standard of 7 mg/m^3 of respirable dust [13]. This standard was successfully introduced to the UK coalfields and its application was associated with a much reduced prevalence of CWP. The USA took the same data and developed their own standard of 2 mg/m^3 respirable dust (measured differently from the UK) which also produced enormous advantages in worker protection [14].

Throughout the middle and later twentieth century there were similar attempts to determine safe levels of asbestos and quartz, with varying success in terms of getting the airborne levels lowered by adoption of tighter standards over time This was often exacerbated by resistance from industries to admitting the harmfulness of the dusts and their reluctance to invest in costly dust suppression equipment.

THE RISE OF PARTICLE TOXICOLOGY

Case reports and epidemiology usually have made the first association between exposure to a dust and development of disease, and epidemiology can explore the exposure-response relationship that is so necessary for standard setting. However toxicology can also make a valuable contribution to prove that an exposure is indeed the cause of a disease using Koch's postulates type of arguments. Additionally, as discussed later it can be used to analyse the properties of particles that make them harmful, something that is difficult to do in the mixed exposures in human populations, and can thus provide information on mechanisms. E.J. King can justifiably be seen as the first major UK particle toxicologist, although he was a Canadian by birth [15]. Between the middle 1940s and the early 1960s he systematically demonstrated silicosis in rats exposed to quartz and mixed dusts [16–18], its modification with aluminium compounds [19], the effects of asbestos on rat lungs [20] and the effects of particles on cells in culture [21]. Early theories of the mechanisms of quartz toxicity divided into 2 camps—the solubility theory and the mechanical theory. It is interesting to note that the Canadian group of Denny and co-workers, in pursuit of the solubility theory, had demonstrated a protective effect of aluminium on silicosis in rabbits in the late nineteen thirties [22] (Figure 5).

Figure 5. (Left) Earl Judson King (1901–1963) and (right) a figure from his 1946 paper**[20]**of a fibrotic lung lesion in a rat exposed to asbestos fibre clearly showing a ferruginous body (a coated asbestos fibre; arrow) in the lesion and multinucleate giant cells. (Used with permission J. Clin. Pathol).

The solubility theory was based on the idea that soluble silicic acid released from the quartz surface was the harmful component, whilst the mechanical theory implicated the sharp and angular quartz particles as mechanically irritating the lungs [23]; neither of these paradigms has survived to present day particle toxicology.

The first real mechanistic particle toxicology work on asbestos had an emphasis on the quantification of dose, a vital step, and was carried out by Dr Chris Wagner. Whilst working as a young pathologist in Johannesburg, Wagner had noted pleural mesothelioma, a very unusual tumour, in workers from the crocidolite mining area in the Cape Province of SA and reported this in 1960 [24]; this established the first clear link between asbestos exposure and its hallmark tumour, though earlier reports had noted a connection. He then came to the UK to work in the MRC pneumoconiosis unit in Llandough, Wales. By 1965 Wagner began publishing an influential series of papers beginning with asbestos deposition and retention, the first emphasis on the modern concept of particle dosimetry [25]. He went on to demonstrate mesothelioma production in rats after exposure to crocidolite [26] and then carry out a number of long term rat inhalation studies that set a new standard for this type of study [27–30] exploring the role of asbestos types and their mechanisms. Similar work was carried out in the Institute of Occupational Medicine in Edinburgh led by Dr John M.G. Davis from the mid-seventies onwards. The Davis group went on to produce a sequence of important papers on the relationship between asbestos characteristics and pathogenicity [31–34], many of which were inspired and executed by Rob Bolton, whose PhD had been supervised by John Davis in the early seventies. In the USA at the same time, Merle Stanton at the National Cancer Institutes was studying mesothelioma induction by asbestos in rats using an implantation technique onto the visceral pleura. The Wagner, Davis and Stanton groups all identified critical roles for fibre type and length, igniting a substantial amount of research on the role played by length, fibre type and biopersistence in toxic potency that continues to the present day and which has defined the first real structure:toxicity model for a pathogenic dust [35–37]. In addition this work was important from a practical point of view in terms of defining exposure, since it allowed formulation of the fibre-counting rules and led to the Walton Beckett graticule [38] which assists in the identification of respirable fibres. This enabled a scientific method of fibre counting and thus regulation of the exposure to fibres in workplace air.

Following on from Stanton's work, in Germany Friedrich Pott and co-workers championed the use of the intraperitoneal test as a sensitive assay for the potential of fibres to cause mesothelioma. The use of the intraperitoneal test has, however, proved a controversial issue, although it has become established in European regulation (Figure 6).

Figure 6. Left Dr Chris Wagner (1923–2000) MRC Pneumoconiosis Research Unit, Llandough; (From the website of the Mesothelioma Interest Group http://imig.org/about/wagner-award-recipients-2 **).** Right Dr John M.G. Davis (1935–2011) Institute of Occupational Medicine, Edinburgh. Property of KD)

PARTICLE TOXICOLOGY COMES OF AGE

By the late seventies the pace in particle toxicology was quickening and there was a burgeoning community of particle toxicologists, heralding the dawn of cellular toxicology and applying its techniques to cells in culture as the new methods became available. The first in the series of meetings, still ongoing, then named 'In vitro effects of mineral dusts' was held in 1979 (see [5] for the history of this meeting). This era witnessed the first attempts to find predictive tests for toxic particles, although the understanding of the key pathological processes such as fibrosis and cancer was not sufficiently advanced for assays to be soundly based on an understanding of the molecular pathology. In fact the Proceedings of the first meeting, held in Cardiff in 1979 [39], were dominated by papers measuring cell death, since that was about all that could be quantified! A few standard methods came into use that still operate now—e.g. the use of a control, harmless particle like TiO2 and a positive particle like quartz to allow benchmarking of any unknown particles under test. Organ culture was also used and [Brooke T. Mossman 40] in the University of Vermont used hamster trachea to demonstrate the effects of asbestos and went on to investigate the molecular mechanism involved in asbestos effects over the next 30 years, pioneering the oxidative stress theory and demonstrating the molecular pathways by which particles influence cells. By the early eighties Arnold Brody in the National Institute of Environmental Health Sciences in North Carolina was using novel and creative electron microscopy approaches to open up the 'black box' of the lung with regard to the early responses to pathogenic dust inhalation. Brody demonstrated that there was high efficiency deposition of inhaled respirable fibres in the centri-acinar and alveolar duct regions of rats [41, 42] leading to high localised dose and a rapid pathobiological response [43, 44]. David Warheit, later of DuPont, worked in NIEHS at this time and was involved in these ground-breaking studies [45], which were the first to demonstrate the rapidly evolving response to a pathogenic dust in this sensitive area, identifying the mechanism of macrophage recruitment to sites of fibre deposition, followed by proliferation of epithelial and other cells and prototypic fibrotic lesion development at the first alveolar duct bifurcation.

In the meantime, in the eighties and nineties new particles came along to occupy particle toxicologists as the 'big three' became better understood. These were the man-made mineral fibres (MMMF), later renamed man-made vitreous fibres (MMVF) and even later renamed synthetic vitreous fibres (SVF). Some of the SVF were seen as direct replacements for banned asbestos, to fill the gap in the industrial need for fibres, but were still fibres and so suspicion arose that asbestos might be being replaced by a similar hazard. These SVF, glasses and ceramics of various sorts that are spun or blown out into fibres, are used for insulation and numerous other purposes. Many studies carried out comparisons between asbestos and the vitreous fibres, culminating in the emergence of 'biopersistence' as a key explanatory attribute for discriminating between carcinogenic and non-carcinogenic long fibres [46, 47]. This era culminated in the IARC Working party on the carcinogenesis of these materials in 2001 [48]. The outcome of these deliberations was that biopersistence dominated; the non-biopersistent SVF were exonerated as carcinogens and the more biopersistent ones were classified as carcinogens. A side-effect of the discovery of biopersistence as a key factor in SVF pathogenicity was the uncovering of biopersistence as the explanation for differences between the more harmful amphibole asbestos and the less harmful chrysotile asbestos, the latter having been shown to be less persistent both *in vitro* and *in vivo*. This period showed the first full application of a number of tests in particle toxicology that are now standard, including genotoxicity tests such as the micronucleus assay and the rise of the measurement of cytokines and use of bronchoalveolar lavage to study lung inflammation and understand mechanisms of inflammation development. It also witnessed the development of a benchmark standard that still pertains today for particle/fibre inhalation studies in the form of the studies by the Research and Consulting Company (RCC) on SVF [49, 50]. These emphasised exposure to respirable diameter fibres of the length found in workplaces, lung burden studies, non-destructive aerosolisation, accurate bimodal fibre counting, bronchoalveolar lavage (BAL) etc. Similar but smaller—scale inhalation studies performed in Edinburgh under the auspices of the Colt Fibre Research Programme in the Institute of Occupational Medicine in the early nineties came to the same conclusion regarding the central role of biopersistence in fibre toxicity [51–53].

AMBIENT PARTICLES

By the middle nineties, with SVF receiving less research attention in the wake of the IARC rulings, environmental particles (PM_{10}) became of increasing interest for particle toxicologists. The prevention of coal burning in urban areas by the enactment of the Clean Air Acts in the UK and elsewhere, had largely eliminated the dense winter smoke fogs that saw epidemics of thousands of deaths, as in the 'great smog' of 1952 in London, and there was a feeling that the problem of pollution in cities had been largely solved. However, epidemiological studies of huge populations, entire cities, began to show correlations between death rates and concentrations of particles and sulphur dioxide in the air. The studies revealed a short-term relationship between these two variables in 'time-series' studies and more long-term effects of living in polluted areas in longitudinal studies. The most famous of the latter type was the study which examined six US cities that had differe nt average levels of particulate air pollution and showed that death rates in each city were related to the concentrations of the particulate air pollution [54]. It seemed that, although the visible soot-laden air that had characterised cities since earliest times was no longer evident, the increase in traffic had brought a new kind of air pollution derived largely from combustion in vehicle engines.

By the early 1990s it was apparent that although the mass of particles in air pollution had fallen dramatically in industrial societies in the West, effects on health in terms of heart and lung disease were still apparent, though by now the predominant source of the pollution had changed from coal burning to vehicles. The introduction of new statistical techniques made it possible to demonstrate that relatively small fluctuations in pollution concentrations were associated with similarly small fluctuations in health effects, from heart attack deaths to consultations with doctors for exacerbations of asthma. Whether these effects derived from particles or toxic gases was a major issue, particularly in terms of what to regulate, but the most remarkable fact was the low concentrations at which the associations were observable. It was notable that although as one might expect respiratory illness occurred, the greater number of deaths and hospitalisations occurred from heart disease, usually heart attack [55]. This fact, that an organ distant from that in which the toxic agent was deposited was the one most affected seemed paradoxical and was an obvious challenge to toxicology. As research advanced, particles were seen to act by a variety of pathways to induce inflammation and oxidative stress in the lungs [56–58], making the link to exacerbations of airways disease and even possibly lung cancer, though the epidemiology on this was, at the time, weak. The question as to how pulmonary deposition of extremely low concentrations of mainly carbon particles might cause heart attack was first addressed in a hypothesis paper that emanated from a discussion in 1994 between the authors, David Godden and Bill MacNee. Seaton, following a conversation with Robert Waller in 1993, chanced upon a paper on seasonal plasma fibrinogen fluctuations and it immediately occurred to him that pollution exposure might act through such an amplifying mechanism that might lead to cardiovascular deaths by promoting clotting in the coronary arteries. The question of low dose was addressed by thinking in terms of numbers rather than mass. It had been known since the 19th century that air pollution contained many very small particles, and this was demonstrated by the Lawther and Waller group in the 1960s. The highly original work of Ferin and Oberdorster and colleagues had shown unexpected increases in toxicity of ultrafine particles, or what are now generally characterised as nanoparticles [59]. Thus we were able to propose that '*...inflammation* *provoked by ultra-fine particles, in addition to promoting exacerbations of lung disease, has an additional effect on the coagulability of blood, increasing the susceptibility of individuals to acute episodes of cardiovascular disease...'* [55]. This brief but highly–cited paper preceded a huge research effort and clear evidence of a link from the lungs to the blood vessel wall and the atherosclerotic plaque subsequently emerged in many studies, for example using ApoE mice [60] and human chamber studies [61]. In fact a persuasive mechanism has emerged based on relatively low exposures of human subjects to diesel exhaust particles (DEP) inducing lung inflammation and oxidative stress. These indirectly affect cardiac blood flow, [62] the associated endothelium [63, 64] and the clotting system [65] in ways that favour atherothrombosis. These data have emerged from a fruitful collaboration between Edinburgh University (Professor David Newby, Dr Nick Mills), University of Umea in Sweden (Professor Thomas Sandström, Professor Anders Blomberg) and the RIVM in the Netherlands (Professor Flemming Cassee). The overall approach utilises cells, rodent and human studies in a highly innovative model program aimed at determining the link between pulmonary deposition of diesel particles and atherothrombosis. Whilst indirect effects of inflammatory mediators and oxidants from the lung means that a role for translocation of the combustion-derived nanoparticles to the blood in driving these events seems to be receding, translocation still remains an open question.

This emphasis on diesel exhaust nanoparticles is not meant to detract from the likelihood that the fine and coarse fractions may have some role in the adverse effects of PM depending on the origin of the particles and their chemical composition e.g. [66].

NANOPARTICLE TOXICOLOGY /NANOTOXICOLOGY

The 1990s saw extraordinary advances in technology which gave rise to the possibilities of engineering materials at sub-micron scale and thereby producing new materials with specifically enhanced or altered properties, nanotechnology. It was foreseen that this would open up a whole new world of industrial and commercial applications, but at the same time concerns began to be expressed that some of these altered properties might imply altered or enhanced hazard to humans or the natural environment. The UK government asked the Royal Academy of

Engineering and the Royal Society to consider these opportunities and uncertainties. At the same time the term nanotoxicology was coined in a commentary by Robert Service in the journal Science in 2004 [67] and a number of European particle toxicologists, led by one of us (KD), laid out what amounted to a manifesto on this new area of particle toxicology in an Editorial in the UK journal Occupational and Environmental Medicine entitled 'Nanotoxicology: A new frontier in particle toxicology relevant to both the workplace and general environment and to consumer safety' [68]. This concern about new manufactured nanoparticles (MNP) being developed in industry was predicated on the experience of particle toxicology which had shown that surface area and reactivity were important determinants of toxicity and which had suggested that combustion-derived nanoparticles were a main factor in ambient particle toxicity (see [4]for a fuller history and the origins of nanotoxicology). The possibility that new nanotechnologies might produce other materials posing similar hazards to asbestos or air pollution particles was a powerful driver in raising the concerns that were crystallised in the influential report of the UK Royal Academy of Engineering and Royal Society [69] (Figure 7).

There has been an enormous amount written on the issue of nanotoxicology, much of it gėnerating more heat than light, but it remains at the time of writing the primary concern for particle toxicologists, and the primary source of funding. Carbon nanotubes for example have been found to be highly fibrogenic in the lungs [70] and if they are in a fibrous morphology, to have the potential to produce asbestos-like effects in the peritoneal and pleural spaces [71, 72]. These toxicological findings have already influenced risk assessment and standard setting in relevant workplaces.

It is interesting to observe that prior to the development of manufactured nanoparticles (MNP), concern that a particle could have an adverse effect tended to develop in the following way:

1. a single or a few cases were reported in a workforce who handled and were exposed to a specific particle.

2. toxico-pathological studies suggested a causative relationship between the material inhaled and disease.

3. epidemiological investigations demonstrated the nature of the relationship and possibly estimated an exposure-response relationship.

4. toxicologists confirmed and further enhanced understanding by controlled animal experimentation and explored the molecular mechanism using cell models.
5. particle toxicologists examined the role of the particle structure in the toxicity.

Figure 7. Royal Society and Royal Academy of Engineering report 'Nanoscience and nanotechnologies: opportunities and uncertainties' published in 2004.

It is striking however, that for MNP this process is working the other way round. Many MNP under study are in general not yet being handled in large enough quantities or for long enough for disease to emerge. Therefore toxicologists are trying to predict the diseases or pathological reactions that might arise from exposure to MNP before there is any evidence that they are occurring or will occur. This would seem to be an improvement on waiting to 'count bodies' on the introduction of a new hazard. However, it does raise the possibility of false positives especially if significant exposures do not occur. This is new territory and the predictive power of the assays we use are being themselves tested to the breaking point while exposure data need to be integrated with the toxicological data to decide risk—not an easy process. It remains to be seen how this will all resolve and which if any nanomaterials represent significant risk.

CIGARETTE SMOKE

Cigarette smoke certainly exceeds 'the big three' in terms of the ill health that it has caused but, apart from a smattering of papers in the early years, particle toxicology has not embraced cigarette smoke. This has become even more marked in the last 20 years, when the ability to publish toxicological (as opposed to epidemiological) research on cigarette smoke effects has been greatly impaired and those studying cigarette smoke, or taking funds for that purpose, have become to an extent marginalised by the scientific community. To our knowledge no papers on cigarette smoke have ever appeared in the journals Particle and Fibre Toxicology nor in Nanotoxicology and very few have ever appeared in the Inhaled Particles series or the Particle Toxicology series. From the point of view of toxicology the relationship between the main pathological consequences in the lungs—cancer and COPD—and particulate versus the gaseous and organic phases is not resolved and particles themselves may not play the primary role. Cigarette smoke remains a marginal issue for particle toxicologists and one that has complicated political and social issues related to it.

THE DOSE IN PARTICLE TOXICOLOGY

It seems important to finish on 'dose', the most pertinent entity in toxicology. In particle toxicology, good dosimetry and the 'biologically effective dose' have proved elusive. Tissue burdens, including mass balance toxicokinetics, are necessary in order to understand the complete picture of any toxin, especially the role of the translocated dose that reaches beyond the portal of entry and its mechanism and consequences. This was seldom done in an integrated way in particle toxicology until the emergence of a key German/USA collaboration. This collaboration, between Gunter Oberdörster and Wolfgang Kreyling has been central to the current view of particle dosimetry that dominates the discipline. Oberdörster worked in the Fraunhöfer in Hanover but relocated to the University of Rochester in the 1980s where there was already strength in depth in particle toxicology through the work of Juraj Ferin and Paul Morrow. Kreyling was, and still is, employed at the Helmholtz near Munich with a large lung dosimetry group which originally included particle toxicologists such as Helmut Greim and Joachim Heyder. Kreyling's strength in mass balance particle toxicokinetics was matched by Oberdörster's insightful and skilful command of toxicology and between the two of them they virtually invented modern particle dosimetry.

This has an emphasis on quantitative aspects of particle dose including the dose rate, the need to understand deposition, clearance and translocation and their relation to pathobiological mechanisms and the response seen at the tissue and cell level, in the lungs and beyond.

Postscript

The era of particle toxicologists who originally saw coalmine dust, quartz and asbestos as the big problem and were involved in the invention of particle toxicology as a discipline is almost over. Nanotoxicology now dominates particle toxicology and for many people nanoparticle toxicology isparticle toxicology. In fact the issues surrounding the big three have never been fully resolved and there remain many unanswered questions in conventional particle toxicology. For example, why is erionite so potent a cause of mesothelioma? What is the mechanism of quartz toxicity? Why do only rats get overload? The avalanche of funding that flowed into nanotoxicology will end soon, new particles and new issues will arise, and many questions remain in conventional particle toxicology, so there will always be a place for particle toxicologists. Multidisciplinarity has always been a feature of particle toxicology and this has been enhanced with the rise of nanotoxicology. The emphasis placed on particle characterisation in nanotoxicology has increased the need for particle toxicologists to work with chemists, physicists and material scientists. It is hoped that the outcome of this is that we edge nearer to a structure-toxicity relationship, the "philosopher's stone" of particle toxicology. The realisation of a full structure-toxicity model would mean that determination of the physicochemical properties of a new particle would enable its toxicity to be predicted without recourse to animal or cell experiments.

While particle toxicology remains a niche area it now has its own specific journals—Particle and Fibre Toxicology and Nanotoxicology; other nanomaterial journals also accept papers on nanotoxicology whilst particle toxicology papers occasionally make it into the general molecular medicine literature. It is to be hoped that having our own journals helps our general profile and that the Impact Factors for these journals continues to increase.

Those new to particle toxicology are encouraged to consult the series of Proceedings of the rolling meetings on particle toxicology that is mentioned earlier in this review (see review [5]) to gain more detailed insight into how particle toxicology has developed, if they are interested. If you are reading this then you are probably contributing to what will become the history of particle toxicology in the future and we hope that this short history has helped to place the subject in context.

AUTHORS' CONTRIBUTIONS

KD and AS contributed equally to the manuscript, drawing on their experience of the history of particle toxicology. Both authors read and approved the final manuscript.

REFERENCES

1. Aitken R, Borm PJA, Ichihara G, Loft S, Marano F, Maynard AD, *et al.*: **Nanoparticles: one word, a multiplicity of hazards (Editorial).** *Nanotoxicology* 2009, **3:** 263–264

2. Hunter D, *et al.*: **The Diseases of Occupations.** *English Universities Press* 1995, 1–1100.

3. Morgan WKC, Seaton A: *Occupational Lung Diseases*. Saunders, Philadelphia; 1995.

4. Oberdorster G, Stone V, Donaldson K: **Toxicology of nanoparticles: A historical perspective.** *Nanotoxicology*2007, **1:** 2–25.

5. Donaldson K, Borm PJA: **Particle paradigms.** *Inhal Toxicol* 2000,**12**(Suppl 3):1–6.

6. Cummins SL: **PNEUMONOCONIOSIS: WITH SPECIAL REFERENCE TO THE SILICO-ANTHRACOSIS OF COAL MINERS.** *Br Med J* 1935, **2:** 287–290.

7. GOUGH J: **The pathology of pneumoconiosis.** *Postgrad Med J* 1949, **25:** 611–618. illust

8. Thomas CR, Kelley TR: **A brief review of silicosis in the United States.** *Environ Health Insights* 2010, **4:** 21–26.

9. Cherniack M: *The Hawk's Nest Incident*. Yale University Press, New Haven; 1986:1–187.

10. Donaldson K, Murphy F, Duffin R, Poland C: **Asbestos, carbon nanotubes and the pleural mesothelium: a review of the hypothesis regarding the role of long fibre retention in the parietal pleura, inflammation and mesothelioma.** *Particle and Fibre Toxicology* 2010.

11. Hodgson JT, McElvenny DM, Darnton AJ, Price MJ, Peto J: **The expected burden of mesothelioma mortality in Great Britain from 2002 to 20503.** *Br J Cancer* 2005, **92:** 587–593.

12. Morgan WKC, Seaton A (Eds): *Occupational Lung Diseases. 3rd Edition, WB Saunders, Philadelphia*. Philadelphia: WB Saunders; 1995.

13. Jacobsen M, Rae S, Walton WH, Rogan JM: **New dust standards for British coal mines.** *Nature* 1970, **227:**445–447.

14. Attfield MD, Kuempel ED: **Pneumoconiosis, coalmine dust and the PFR.** *Ann Occup Hyg* 2003, **47:** 525–529.

15. Signy AG: **Earl Judson KING.** *J Clin Pathol* 1963, **16:** 92–93.

16. King EJ, Rogers N: **The effect of olivine on the lungs of rats.** *J Pathol Bacteriol* 1945, **57:** 488–491.

17. King EJ, Ray SC: **The effects of granite on the lungs of rats.** *Br J Ind Med* 1950, **7:** 37–41.

18. Ray SC, King EJ, Harrison CV: **The action of variable amounts of quartz on the lungs of rats; the extent of pathological change in relation to the amount injected.** *Br J Ind Med* 1951, **8:** 62–67.

19. King EJ, Harrison CV, Mohanty GP, Yoganathan M: **The effect of aluminium and of aluminium containing 5 per cent. of quartz in the lungs of rats.** *J Pathol Bacteriol* 1958, **75:** 429–434.

20. King EJ, Clegg JW, RAE VM: **The effect of asbestos, and of asbestos and aluminium, on the lungs of rabbits.** *Thorax* 1946, **1:** 188–197.

21. Belt TH, Friedmann I, KING EJ: **The effect of asbestos on tissue cultures; a comparative study with quartz and coal dust.** *J Pathol Bacteriol* 1947, **59:** 159–164.

22. Denny JJ, Robson WD, Irwin DA: **The prevention of silicosis by metallic Aluminum ii.** *Can Med Assoc J* 1939,**40:** 213–228.

23. Cooke WE: **Modern views on silicosis.** *J Hyg (Lond)* 1935, **35:** 207–218.

24. Wagner JC, SLEGGS CA, MARCHAND P: **Diffuse pleural mesothelioma and asbestos exposure in the North Western Cape Province 138.** *Br J Ind Med* 1960, **17:** 260–271.

25. Wagner JC, Skidmore JW: **Asbestos dust deposition and retention in rats.** *Ann N Y Acad Sci* 1965, **132:** 77–86.

26. Wagner JC, Berry G: **Mesotheliomas in rats following inoculation with asbestos.** *Br J Cancer* 1969, **23:** 567–581.

27. Wagner JC, Berry G, Skidmore JW, Timbrell V: **The effects of the inhalation of asbestos in rats.** *Br J Cancer*1974, **29:** 252–269.

28. Wagner JC: **Tumours in experimental animals following exposure to asbestos dust.** *Ann Anat Pathol (Paris)*1976, **21:** 211–214.

29. Berry G, Wagner JC: **Effect of age at inoculation of asbestos on occurrence of mesotheliomas in rats.** *Int J Cancer* 1976, **17:** 477–483.

30. Morgan A, Davies P, Wagner JC, Berry G, Holmes A: **The biological effects of magnesium-leached chrysotile asbestos.** *Br J Exp Pathol* 1977, **58:** 465–473.

31. Davis JM, Beckett ST, Bolton RE, Donaldson K: **A comparison of the pathological effects in rats of the UICC reference samples of amosite and chrysotile with those of amosite and chrysotile collected from the factory environment.** *Biological Effects of Mineral Fibres, IARC Scientific Publication No 30 Lyon, France, 1980* 1980, 258–292.

32. Bolton RE, Davis JM, Donaldson K, Wright A: **Variations in the carcinogenicity of mineral fibres 24.** *Ann Occup Hyg* 1982, **26:** 569–582.

33. Davis JG, Addison J, Bolton RE, Donaldson K, Jones AD, Miller BG: **Inhalation studies on the effects of tremolite and brucite dust in rats.** *Carcinogenesis* 1985, **6:** 667–674.

34. Davis JG, Addison J, Bolton RE, Donaldson K, Jones AD, Smith T: **The pathogenicity of long versus short fiber samples of amosite asbestos administered to rats by inhalation and intraperitoneal injection.** *Br J Exp Pathol* 1986, **67:** 415–430.

35. Stanton MF, Wrench C: **Mechanisms of mesothelioma induction with asbestos and fibrous glass.** *J Natl Cancer Inst* 1972, **48:** 797–821.

36. Stanton MF, Laynard M, Tegeris A, Miller E, May M, Kent : **Carcinogenicity of fibrous glass: pleural response in the rat in relation to fiber dimension.** *J Natl Cancer Inst* 1977, **58:** 587–603.

37. Stanton MF, Layard M, Tegeris A, Miller E, May M, Morgan E, *et al.*: **Relation of particle dimension to carcinogenicity in amphibole asbestoses and other fibrous minerals.** *J Natl Cancer Inst* 1981, **67:** 965–975.

38. Walton WH, Beckett ST: **A microscope eyepiece graticule for the evaluation of fibrous dusts.** *Ann Occup Hyg* 1977, **20:** 19–23.

39. Brown RC, Gormley IP, Chamberlain M, Davies R (Eds): *The in vitro effects of mineral dusts*. Academic Press, London; 1980.

40. Mossman BT, Craighead JE: **Use of hamster tracheal organ cultures for assessing the cocarcinogenic effects of inorganic particulates on the respiratory epithelium.** *Prog Exp Tumor Res* 1979, **24:** 37–47.

41. Warheit DB, Chang LY, Hill LH, Hook GR, Crapo JD, Brody AR: **Pulmonary macrophage accumulation and asbestos-induced lesions at sites of fiber deposition.** *Am Rev Respir Dis* 1984, **129:** 301–310.

42. Brody AR, Warheit DB, Chang LY, Roe MW, George G, Hill LH: **Initial deposition pattern of inhaled minerals and consequent pathogenic events at the alveolar level.** *Ann NY Acad Sci* 1984, **428:** 108–120.

43. Brody AR, Hill LH, Warheit DB: **Induction of early alveolar injury by inhaled asbestos and silica.** *Fed Proc*1985, **44:** 2596–2601.

44. Brody AR, Liu JY, Brass D, Corti M: **Analyzing the genes and peptide growth factors expressed in lung cells in vivo consequent to asbestos exposure and in vitro.** *Environ Heal Perspect* 1997, **105:** 1165–1171.

45. Warheit DB, George G, Hill LH, Snyderman R, Brody AR: **Inhaled asbestos activates a complement-dependent chemoattractant for macrophages.** *Lab Investig* 1985, **52:** 505–514.

46. Hesterberg TW, Hart GA, Chevalier J, Miiller WC, Hamilton RD, Bauer J, *et al.*: **The importance of fiber biopersistence and lung dose in determining the chronic inhalation effects of X607, RCF1, and chrysotile asbestos in rats.** *Toxicol Appl Pharmacol* 1998, **153:** 68–82.

47. McDonald JC: **Mineral fibre persistence and carcinogenicity.** *Ind Health* 1998, **36:** 372–375.

48. WHO IARC: *IARC monogrpahs on the evaluation of carcinogenic risks to humans. Volume 81 Man-made vitreous fibres*. IARC Lyon; 2002

49. Bunn WB, Bender JR, Hesterberg TW, Chase GR, Konzen JL: **Recent studies of man-made vitreous fibers—chronic animal inhalation studies.** *J Occup Med* 1993, **35:** 101–113.

50. Hesterberg TW, Miller WC, Mcconnell EE, Chevalier J, Hadley JG, Bernstein DM, *et al.*: **Chronic inhalation toxicity of size-**

separated glass-fibers in fischer-344 rats. *Fundam Appl Toxicol* 1993, **20:** 464–476.

51. Miller BG, Searl A, Davis JM, Donaldson K, Cullen RT, Bolton RE, *et al.*: **Influence of fibre length, dissolution and biopersistence on the production of mesothelioma in the rat peritoneal cavity.** *Ann Occup Hyg* 1999,**43:** 155–166.

52. Searl A, Buchanan D, Cullen RT, Jones AD, Miller BG, Soutar CA: **Biopersistence and durability of nine mineral fibre types in rat lungs over 12 months.** *Ann Occup Hyg* 1999, **43:** 143–153.

53. Miller BG, Jones AD, Searl A, Buchanan D, Cullen RT, Soutar CA, *et al.*: **Influence of characteristics of inhaled fibres on development of tumours in the rat lung.** *Ann Occup Hyg* 1999, **43:** 167–179.

54. Dockery DW, Pope CA, Xu XP, Spengler JD, Ware JH, Fay ME, *et al.*: **An association between air-pollution and mortality in 6 united-states cities.** *N Engl J Med* 1993, **329:** 1753–1759.

55. Seaton A, MacNee W, Donaldson K, Godden D: **Particulate air-pollution and acute health-effects.** *Lancet*1995, **345:** 176–178.

56. Ghio AJ, Devlin RB: **Inflammatory lung injury after bronchial instillation of air pollution particles.** *Am J Respir Crit Care Med* 2001, **164:** 704–708.

57. Molinelli AR, Madden MC, Mcgee JK, Stonehuerner JG, Ghio AJ: **Effect of metal removal on the toxicity of airborne particulate matter from the Utah Valley.** *Inhal Toxicol* 2002, **14:** 1069–1086.

58. Jimenez LA, Drost EM, Gilmour PS, Rahman I, Antonicelli F, Ritchie H, *et al.*: **PM(10)-exposed macrophages stimulate a proinflammatory response in lung epithelial cells via TNF-alpha.** *Am J Physiol Lung Cell Mol Physiol* 2002, **282:** L237-L248.

59. Ferin J, Oberdorster G, Penney DP: **Pulmonary retention of ultrafine and fine particles in rats.** *Am J Respir Cell Mol Biol* 1992, **6:** 535–542.

60. Lippmann M, Hwang JS, Maciejczyk P, Chen LC: **PM source apportionment for short-term cardiac function changes in ApoE−/− mice 5.** *Environ Health Perspect* 2005, **113:** 1575–1579.

61. Pekkanen J, Peters A, Hoek G, Tiittanen P, Brunekreef B, de Hartog J, *et al.*: **Particulate air pollution and risk of ST-segment depression during repeated submaximal exercise tests among subjects with coronary heart disease: the Exposure and Risk Assessment for Fine and Ultrafine Particles in Ambient Air (ULTRA) study.** *Circulation* 2002, **106:** 933–938.

62. Cruts B, van Etten L, Tornqvist H, Blomberg A, Sandstrom T, Mills NL, *et al.*: **Exposure to diesel exhaust induces changes in EEG in human volunteers.** *Part Fibre Toxicol* 2008, **5:** 4.

63. Tornqvist H, Mills NL, Gonzalez M, Miller MR, Robinson SD, Megson IL, *et al.*: **Persistent endothelial dysfunction in humans after diesel exhaust inhalation.** *Am J Respir Crit Care Med* 2007, **176:** 395–400.

64. Shaw CA, Robertson S, Miller MR, Duffin R, Tabor CM, Donaldson K, *et al.*: **Diesel Particulate-Exposed Macrophages Cause Marked Endothelial Cell Activation.** *Am J Respir Cell Mol Biol* 2011, **44:** 840–851.

65. Lucking AJ, Lundback M, Mills NL, Faratian D, Barath SL, Pourazar J, *et al.*: **Diesel exhaust inhalation increases thrombus formation in man.** *Eur Heart J* 2008, **29:** 3043–3051.

66. Steenhof M, Gosens I, Strak M, Godri KJ, Hoek G, Cassee FR, *et al.*: **In vitro toxicity of particulate matter (PM) collected at different sites in the Netherlands is associated with PM composition, size fraction and oxidative potential–the RAPTES project.** *Part Fibre Toxicol* 2011, **8:** 26.

67. Service RF: **Nanotoxicology. Nanotechnology grows up.** *Science* 2004, **304:** 1732–1734.

68. Donaldson K, Stone V, Tran CL, Kreyling W, Borm PJ: **Nanotoxicology.** *Occup Environ Med* 2004, **61:** 727–728.

69. Royal Society and Royal Academy of Engineering: **Nanoscience and nanotechnologies: opportunities and uncertainties.** *The Royal Society. London 2004,* 116pp.

70. Shvedova AA, Kisin ER, Murray AR, Johnson VJ, Gorelik O, Arepalli S, *et al.*: **Inhalalation versus aspiration of single walled carbon nanotubes in C57/Bl6 mice: Inflammation, fibrosis,**

oxidative stress and mutagenesis. *Am J Physiol Lung Cell Mol Physiol* 2008, **295:** L552-L565.

71. Poland CA, Duffin R, Kinloch I, Maynard A, Wallace WA, Seaton A, *et al.*: **Carbon nanotubes introduced into the abdominal cavity of mice show asbestos-like pathogenicity in a pilot study.** *Nat Nanotechnol* 2008, **3:**423–428.

72. Murphy FA, Poland CA, Duffin R, Al Jamal KT, Ali-Boucetta H, Nunes A, *et al.*: **Length-dependent retention of carbon nanotubes in the pleural space of mice initiates sustained inflammation and progressive fibrosis on the parietal pleura.** *Am J Pathol* 2011, **178:** 2587–2600.

Chapter 4

Diesel Exhaust in Miners Study: How to Understand the Findings?

Peter Morfeld[1,2]

[1]Institute for Occupational Epidemiology and Risk Assessment of Evonik Industries, Essen, Germany

[2]Institute and Policlinic for Occupational Medicine, Environmental Medicine and Prevention Research, University of Cologne, Cologne, Germany

ABSTRACT

The Diesel Exhaust in Miners Study (DEMS) is an outstanding epidemiological project on the association between occupational diesel exhaust exposures, measured as long-term respirable elemental carbon (REC) estimates, and lung cancer mortality in a large cohort of US miners. Two articles published recently (Attfield et al. (J Natl Cancer Inst Epub, [2012]), Silverman et al. (J Natl Cancer Inst Epub, [2012])) dsescribed the epidemiological findings. These papers are expected to have considerable impact on the evaluation of the carcinogenic potential of diesel exhaust and, furthermore, on occupational and environmental limit value discussions related to diesel motor emissions and particle exposures. DEMS found remarkable exposure-response relationships between REC exposure estimates and lung cancer mortality - conditional on a pronounced effect of surface vs. underground work on lung cancer risk. If this risk factor is ignored the estimated REC-lung cancer association is attenuated substantially. The authors relied on this risk factor in their main analyses. However, this factor "surface/underground work" remained unexplained. The factor lead the authors to introduce unusual cross-product terms of location and smoking in adjustment procedures and even caused the authors to hypothesize that high REC exposures are protective against lung cancer excess risks due to smoking. To understand the reliability of these conclusions, we should ask basic questions about the data collection process in DEMS: Did the mortality follow-up procedures suffer from errors like those that affected the NCI formaldehyde cohort study? Are the REC and/or smoking data reliable, and are these data collected/constructed in such a way that the procedures allow valid comparisons between surface and underground workers? Without clarifying the issues raised in this Commentary the Diesel Exhaust in Miners Study remains to be difficult to interpret.

Keywords: Diesel exhaust in miners study; Epidemiology; Lung cancer; Respirable elemental carbon

BACKGROUND

Two risk analyses articles [1, 2] about the US Diesel Exhaust in Miners Study (DEMS), were recently published in the Journal of the National Cancer Institute (JNCI). DEMS is an impressive epidemiological investigation organized by the National Institute for Occupational Safety and Health (NIOSH) and the National Cancer Institute (NCI). The authors have to be congratulated for this outstanding research project on the association between occupational diesel exhaust exposures, measured as long-term respirable elemental carbon (REC) estimates, and lung cancer mortality in a large cohort of US miners. The researchers summarized the main finding of DEMS as follows: "To our knowledge, this is the first report of a statistically significant exposure-response relationship for diesel exposure and lung cancer based on quantitative estimates of historical diesel exposure with adjustment for smoking and other potential confounders" [2]. Thus, these two publications [1, 2] are expected to have considerable impact on the evaluation of the carcinogenic potential of diesel exhaust and, furthermore, on occupational and environmental limit value discussions related to diesel motor emissions and particle exposures.

Although this important study was well designed and performed, some issues may deserve a deeper scientific discussion. In the following I'd like to instigate such discussions while mainly focussing on methodological subtleties within DEMS. However, although subtleties, these methodological aspects may have substantial relevance for a reliable interpretation of the study and also for a justifiable application of study results in future evaluations of diesel exhaust. Because a Letter to The Editor and/or a Commentary on DEMS was not acceptable for the Editor of JNCI if the comments on the NCI formaldehyde cohort mortality study (cp. point 1 below) were not eliminated I decided to withdraw my submission from JNCI and to publish independently.

MAIN TEXT

In the following I will outline seven issues that deserve a deeper scientific discussion and should be addressed by the DEMS researchers.

1) The authors [1] reported an overall lung cancer SMR of 1.26. Internal analyses showed large increases of relative risks (RR) across estimates of REC exposure: Up to a 5fold or even 7fold lung cancer risk at high exposures (see abstracts of [1] and [2]). Such high RR estimates appear to be compatible with an overall SMR = 1.26 only, if there are deficits of lung cancer deaths among the low exposed (unfortunately, this cannot be justified directly because the authors did not report SMRs across REC exposure categories). If so this resembles the situation of formaldehyde epidemiology where an NCI cohort study [3] showed a pronounced increase of leukemia RRs in internal analyses while this increase was mainly based on a deficit of deaths among the low exposed [4]. Later it became clear that the mortality follow-up of the NCI study was relevantly incomplete [5, 6]: NCI researchers had missed about 1000 deaths out of 9500. These deaths were missed proportionately more among the low exposed workers. This differential mortality follow-up error produced a substantial upward bias in risk estimates across exposure estimates [7]. Because the DEMS [1, 2] was based on an evaluation of the US National Death Index Plus and the Social Security Administration death files like the NCI formaldehyde cohort study [3] and appears to have a similar structure of death deficits in the cancer endpoint of interest among the low exposed, the question arises whether the mortality follow-up of the Diesel Exhaust in Miners Study is really complete. Up to now NCI researchers neither explained why they failed in the mortality

follow-up of the formaldehyde cohort and why this error was differential nor did they correct other obviously incomplete investigations [8] although this was urgently asked for [7]. Thus, the basis of US mortality studies applying follow-up procedures like those used in the NCI formaldehyde cohort study [3] remains dubious and this sheds doubt on the Diesel Exhaust in Miners Study also.

2) The DEMS [1, 2] distinguished between "surface only work" and "ever underground work". Table 2 in [1] reported on the lung cancer SMRs taking age, calendar time, gender, race and state into account. These mortality statistics differed unexpectionally between "surface only work" (SMR = 1.33) and "ever underground work" (SMR = 1.21). For the sake of clarity, I fitted a Poisson model to the data and it returned a ratio of 1.1 with a 0.95-CI of 0.83 - 1.46 and p-value of 0.51 [9]. However, the authors [1] reported a remarkably different finding after a further adjustment for REC exposure in the Cox models (all other covariates were identical to those in the SMR analyses): the ratio increased to 1.9 (range: 1.64-2.28, depending on the Cox model specification). It is perplexing that after adjustment for REC exposure estimates such a large risk factor between "surface only work" and "ever underground work" became apparent that went unnoticed without adjustment for REC exposure in the SMR analyses. However, even this remarkably high RR value of about 2 underestimates the strength of the risk factor "surface work vs. underground work" in a model adjusting for REC exposure because the indicator variable was not set up in an optimal way (see point 3). In the case–control study [2] this puzzling risk factor remained and could not be explained by a difference in smoking habits or any other covariate differences between surface and underground workers. It became even more confusing because a significant interaction between smoking effects and location of work was reported: Smoking was described to have a larger effect on lung cancer mortality when working on surface (i.e., a modification of the smoking effect by location and of the location effect by smoking was found). In all final models of the case–control study [2] REC exposure risk estimates were adjusted by such unusual cross-product variables of the two potential confounders whereas the baseline terms smoking and location were excluded from the model equations simultaneously. The coefficients of these models are difficult to understand. In particular, because the

authors [2] missed to present modelling results as usually reported on in epidemiological studies: results after simultaneously adjusting for the potential confounders smoking and location – but without including interaction terms of these potential confounders. Such results were helpful for comparisons. Even more basic questions remained unanswered: what are the ORs for REC exposure after controlling for only those variables used in SMR calculations [2], and what are the ORs for smoking in underground and surface workers without adjustment for REC exposure?

3) The distinction between surface and underground work is obviously of major importance for an understanding of this study. The authors [1] clearly stated that those Cox models that followed their a-priori defined analysis plan did not show any convincing association between REC exposure estimates and lung cancer mortality. After subdividing the cohort into "surface only workers" vs. "ever underground workers" or after adjusting for a "surface only/ever underground work" indicator the Cox models returned pronounced dose–response relationships with REC exposure estimates. However, neither this subdivision nor the indicator variable made use of all information available. According to the description of the exposure assessment process a REC estimate was allocated to every person-year. And because jobs and REC exposures differed between surface and underground work a reliable exposure assessment should take account of workers' location by year (which I assume was done, otherwise the exposure estimates were obviously inaccurate). Thus, the information whether a worker is on surface or underground should have been available for every person-year. Surprisingly, the authors [1, 2] used this information only in part. They derived the variable "ever underground work" and defined it in a time-dependent manner. It had been more natural, easier to interpret and more complete to analyze the time-dependent variable "underground work" instead. Because the distinction between surface and underground work is central to this study the authors should have better used the full information available.

4) The REC exposure risk estimates differed with location [1]: the authors reported a twenty times higher excess risk per ug/m3-y on "surface only" in comparison to "ever underground". After taking logs of exposure the "surface only" REC coefficient was about twice the one calculated for workers "ever underground". The researchers [1] tested the differences of the effect estimates (significant on the

log scale, not significant on the linear scale) but neither reported how they performed the tests nor did they show any details of the results. The usual way to perform such a test is to add an interaction term "REC exposure x location" to the models. It is surprising that the authors [1, 2] did not present such interaction models whereas they introduced interaction terms among covariates in the case–control models [2]. Such interaction analyses on REC and location may inform us whether the REC effect estimates are homogeneous across location and, thus, whether the REC findings of the DEMS can be generalized.

5) The exposure assessment procedures were criticized to suffer from a considerable uncertainty [10]. The authors responded in detail while making clear that they do not believe and never claimed that the approach is without error. They agreed that imprecision exists in their exposure estimates [11]. Thus, the REC exposure intensity values allocated to each miner and person-year are no exact data describing the truth correctly but should be understood as exposure estimates with uncertainties. However, the DMES analyses [1, 2] applied the exposure data as if fixed and without error. A more reliable analysis should cover these exposure uncertainties quantitatively by simulation procedures [12] or Bayesian analyses [13]. Uncertainties related to the covariate data, e.g. smoking habit information, can also be taken into account [14]. These extended and more realistic analyses may help to decide whether the unavoidable uncertainties of the exposure estimates and of the covariate data may have a relevant effect on the study findings or not.

6) In the following I list some minor points/questions, but they are worth mentioning:

   - Was the study censored at an age when death certificates are generally considered to be less reliable, e.g., at an age of 85 y? (e.g., [15])

   - Was the effect of potential confounders on the REC-OR evaluated on a single basis or simultaneously for all potential confounders of interests? (the authors referred to a change in estimate criterion < 10%, [2])

   - Was missing data always analyzed as a data category although this may have lead to distortions? [16]

- How are the smoking habits distributed across facilities? (at the trona mines smoking was prohibited, [2])
- How do the results change if those workers are dropped from analyses who were hired at an age greater than 40 years? (these miners had a higher probability of prior occupational exposures to carcinogens, [1])
- How do the curvilinear relationships and goodness of fit statistics look like if more general analytical procedures are applied like fractional polynomials [17] or spline regression [18] and evaluated by information criteria [19]?

7) Both papers [1, 2] cited and emphasized the German potash miner cohort study [20]. However, a new and different interpretation of this epidemiological investigation is indicated because a more detailed analysis has been performed taking prior exposures in Uranium mining into account [21]. In this updated analysis the OR per mg/m3-y of total carbon exposure was 1.02 (0.95-CI: 0.80-1.31), and high exposed workers showed an unexceptionable OR = 0.98 (0.95-CI: 0.61-1.58) in comparison to low exposed. Excess lung cancer mortalities were only indicated at very high cumulative exposures to diesel motor emissions. An international publication is currently under peer review.

DISCUSSION

In summary, the Diesel Exhaust in Miners Study [1, 2] described remarkable exposure-response relationships between REC estimates and lung cancer mortality - conditional on a pronounced effect of surface vs. underground work on lung cancer risk. However, if this risk factor is ignored the estimated REC-lung cancer association is attenuated substantially. Unfortunately, the factor "surface/underground work" remained unexplained even in the case–control study [2]. It is worth noticing that the authors [1, 2] relied on this risk factor in their main analyses. This lead them to introduce unusual cross-product terms of location and smoking in adjustment procedures [2] and even caused the authors to hypothesize that high diesel motor emissions are protective against lung cancer excess risks due to smoking and vice versa [2]. Although the authors listed mechanistic considerations in favour of this hypothesis such implications are surprising. Thus, to understand the reliability of these conclusions, we should go back and ask more basic questions about the data collection process: Did the mortality follow-up procedures suffer from errors like those that affected the NCI formaldehyde cohort study [3]? Are the REC and/or smoking data reliable, and are these data collected/constructed in such a way that the procedures allow valid comparisons between surface and underground workers and analyses across these groups?

CONCLUSION

The researchers should be congratulated for the impressive DEMS investigation. However, without clarifying the major issues raised in this Commentary the Diesel Exhaust in Miners Study [1, 2] remains to be difficult to interpret.

ACKNOWLEDGEMENTS

I'd like to thank two unknown reviewers for their helpful comments on an earlier version of this manuscript.

AUTHORS' CONTRIBUTIONS

PM drafted the manuscript. No other contributors were involved.

REFERENCES

1. Attfield MD, Schleiff PL, Lubin JH, Blair A, Stewart PA, Vermeulen R, Coble JB, Silverman DT: **The diesel exhaust in miners study: A cohort mortality study with emphasis on lung cancer.** *J Natl Cancer Inst* 2012.
2. Silverman DT, Samanic CM, Lubin JH, Blair AE, Stewart PA, Vermeulen R, Coble JB, Rothman N, Schleiff PL, Travis WD, *et al.*: **The diesel exhaust in miners study: a nested case–control study of lung cancer and diesel exhaust.** *J Natl Cancer Inst* 2012.

3. Hauptmann M, Lubin JH, Stewart PA, Hayes RB, Blair A: **Mortality from lymphohematopoietic malignancies among workers in formaldehyde industries.** *J Natl Cancer Inst* 2003,**95**(21):1615–1623.

4. Marsh GM, Youk AO: **Reevaluation of mortality risks from leukemia in the formaldehyde cohort study of the National Cancer Institute.** *Regul Toxicol Pharmacol* 2004,**40**(2):113–124.

5. Beane Freeman LE, Blair A, Lubin JH, Stewart PA, Hayes RB, Hoover RN, Hauptmann M: **Mortality from lymphohematopoietic malignancies among workers in formaldehyde industries: the national cancer institute cohort.** *J Natl Cancer Inst* 2009,**101**(10):751–761.

6. Beane Freeman LE, Blair A, Lubin JH, Stewart PA, Hayes RB, Hoover RN, Hauptmann M: **Supplementary data.***J Natl Cancer Inst* 2009,**101**(10):1–7.

7. Marsh GM, Youk AO, Morfeld P, Collins JJ, Symons JM: **Incomplete follow-up in the National Cancer Institute's formaldehyde worker study and the impact on subsequent reanalyses and causal evaluations.***Regul Toxicol Pharmacol* 2010, **58:** 233–236.

8. Hauptmann M, Lubin JH, Stewart PA, Hayes RB, Blair A: **Mortality from solid cancers among workers in formaldehyde industries.** *Am J Epidemiol* 2004,**159**(12):1117–1130.

9. StataCorp: **Stata statistical software: release 11.** *College Station, Texas, USA: StataCorp LP; 2011*

10. Borak J, Bunn WB, Chase GR, Hall TA, Head HJ, Hesterberg TW, Sirianni G, Slavin TJ: **Letter to the editor: Comments on the diesel exhaust in miners study.** *Ann Occup Hyg* 2011,**55**(3):339–342. author reply 343–346

11. Stewart PA, Coble JB, Vermeulen R, Blair A, Lubin J, Attfield M, Silverman DT: **Reply to Borak et al.** *Ann Occup Hyg* 2011,**55**(3):343–346.

12. Lash TL, Fink AK: **Semi-automated sensitivity analysis to assess systematic errors in observational data.***Epidemiol* 2003,**14**(4):451–458.

13. Morfeld P, McCunney RJ: **Bayesian bias adjustments of the lung cancer SMR in a cohort of German carbon black production workers.** *J Occup Med Toxicol* 2010,**5**(1):23.

14. Rothman KJ, Greenland S, Lash TL: *Modern epidemiology*. 3rd edition. Lippincott Williams & Wilkins, Philadelphia; 2008.

15. Sorahan T, Hamilton L, van Tongeren M, Gardiner K, Harrington JM: **A cohort mortality study of U.K. carbon black workers, 1951–1996.** *Am J Ind Med* 2001,**39**(2):158–170.

16. Vach W, Blettner M: **Biased estimation of the odds ratio in case–control studies due to the use of ad hoc methods of correcting for missing values for confounding variables.** *Am J Epidemiol* 1991,**134**(8):895–907.

17. Royston P, Altman DG, Sauerbrei W: **Dichotomizing continuous predictors in multiple regression: a bad idea.** *Stat Med* 2006,**25**(1):127–141.

18. Desquilbet L, Mariotti F: **Dose–response analyses using restricted cubic spline functions in public health research.** *Stat Med* 2010,**29**(9):1037–1057.

19. Burnham KP, Anderson DR: *Model Selection and Multimodel Inference. A Practical Information-Theoretic Approach*. 2nd edition. Springer Science + Business Media, LLC, New York; 2002.

20. Neumeyer-Gromen A, Razum O, Kersten N, Seidler A, Zeeb H: **Diesel motor emissions and lung cancer mortality–results of the second follow-up of a cohort study in potash miners.** *Int J Cancer*2009,**124**(8):1900–1906.

21. Möhner M, Kersten N, Gellissen J: **Zum Einfluss von Dieselmotoremissionen auf das Lungenkrebsrisiko - Reanalyse einer Kohortenstudie im Kalibergbau.** *Arbeitsmed Sozialmed Umweltmed* 2012,**47**(3):150.

Chapter 5

Diesel Engine Exhaust and Lung Cancer Risks – Evaluation of the Meta-Analysis By Vermeulen Et Al. 2014

Peter Morfeld[1,2] and Michael Spallek[3,4]

[1]*Institute and Policlinic for Occupational Medicine, Environmental Medicine and Prevention Research of Cologne University, Kerpener Str. 62, 50937 Köln, Germany*

[2]*Institute for Occupational Epidemiology and Risk Assessment (IERA) of Evonik Industries AG, Rellinghauser Str. 1-11, D-45128 Essen, Germany*

[3]*Institute for Occupational, Social and Environmental Medicine of the Goethe University Frankfurt/Main, Theodor-Stern-Kai 7, 60590 Frankfurt am Main, Germany.*

[4]*European Research Group on Environment and Health in the Transport Sector (EUGT e.V.), Fritschestr. 35, 10627 Berlin, Germany.*

ABSTRACT

Background: Vermeulen et al. 2014 published a meta-regression analysis of three relevant epidemiological US studies (Steenland et al. 1998, Garshick et al. 2012, Silverman et al. 2012) that estimated the association between occupational diesel engine exhaust (DEE) exposure and lung cancer mortality. The DEE exposure was measured as cumulative exposure to estimated respirable elemental carbon in µg/m^3-years. Vermeulen et al. 2014 found a statistically significant dose–response association and described elevated lung cancer risks even at very low exposures.

Methods: We performed an extended re-analysis using different modelling approaches (fixed and random effects regression analyses, Greenland/Longnecker method) and explored the impact of varying input data (modified coefficients of Garshick et al. 2012, results from Crump et al. 2015 replacing Silverman et al. 2012, modified analysis of Moehner et al. 2013).

Results: We reproduced the individual and main meta-analytical results of Vermeulen et al. 2014. However, our analysis demonstrated a heterogeneity of the baseline relative risk levels between the three studies. This heterogeneity was reduced after the coefficients of Garshick et al. 2012 were modified while the dose coefficient dropped by an order of magnitude for this study and was far from being significant ($P = 0.6$). A (non-significant) threshold estimate for the cumulative DEE exposure was found at 150 µg/m^3-years when extending the meta-analyses of the three studies by hockey-stick regression modelling (including the modified coefficients for Garshick et al. 2012). The data used by Vermeulen and colleagues led to the highest relative risk estimate across all sensitivity analyses performed. The lowest relative risk estimate was found after exclusion of the explorative study by Steenland et al. 1998 in a meta-regression analysis of Garshick et al. 2012 (modified), Silverman et al. 2012 (modified according to Crump et al. 2015) and Möhner et al. 2013. The meta-coefficient was estimated to be about 10–20 % of the main effect estimate in Vermeulen et al. 2014 in this analysis.

Conclusions: The findings of Vermeulen et al. 2014 should not be used without reservations in any risk assessments. This is particularly true for the low end of the exposure scale.

Keywords: Diesel DEE Lung cancer Epidemiology Meta-analysis

BACKGROUND

Vermeulen et al. [1] published a meta-regression analysis of three major epidemiological US studies [2–4] which investigated the association between diesel engine exhaust (DEE) exposure – based on the cumulative exposure to respirable elemental carbon (REC) in µg/m^3-years – and lung cancer mortality. The authors reported that some of them worked as members of the IARC Working Group (see [5]), which produced a review of the carcinogenic effect of DEE (hazard assessment) and classified DEE as a human carcinogen (IARC Group 1) in 2012. With this follow-up analysis, Vermeulen et al. [1] aimed to continue these considerations and contribute to the quantitative risk assessment based on the major studies. From the data of the three studies, the authors' main result was a common exposure-risk curve without a threshold and concluded: "We estimated a lnRR of 0.00098 (95 % CI: 0.00055, 0.0014) for lung cancer mortality with each 1-$\mu g/m^3$-year increase in cumulative EC based on a linear meta-regression model. Corresponding lnRRs for the individual studies ranged from 0.00061 to 0.0012. Estimated numbers of excess lung cancer deaths through 80 years of age for lifetime occupational exposures of 1, 10, and 25 $\mu g/m^3$ EC were 17, 200, and 689 per 10,000, respectively. For lifetime environmental exposure to 0.8 $\mu g/m^3$ EC, we estimated 21 excess lung cancer deaths per 10,000. Our estimates suggest that stringent occupational and environmental standards for DEE should be set." So the paper describes elevated risks of cancer even at very low DEE exposures. As a result, the Vermeulen et al. [1] paper will play an important role in the ongoing DEE limit value discussions. The US National Institute for Occupational Health (NIOSH) has already attempted to derive estimates from this using its standard procedure (excess case calculation) [6]: Based on Vermeulen et al. [1], Dr. Park calculated an estimate for the 8 h DEE threshold at the workplace of 0.59 $\mu g/m^3$.

In contrast, the EU's ACSH [7] recommended a limit value of 100 $\mu g/m^3$ for DEEE (diesel engine exhaust emissions), measured as elemental carbon. This recommendation is not health-based but reflects mainly socio-economic considerations. Cherrie et al. [8] concluded that only 2 % of workers exposed to DEEE are estimated to be exposed above this level in the EU. In addition, the authors wrote: "There is a case for introducing an OEL [occupational exposure limit] for DEE particulate, but the OEL would need to be much lower than the typical European OEL that we tested (0.1 mg/m^3)". The proposal derived by Park [6] using Vermeulen et al. [1] is such a proposal that is much lower than the recommendation of the EU's ACSH [7]. A critical analysis of the meta-regression approach by Vermeulen et al. [1] is indicated to understand whether the Vermeulen et al. analysis presents evidence to support such a low limit value, far-off the 100 $\mu g/m^3$ limit value recommendation of ACSH [7].

Throughout this paper, we use "dose" as an abbreviation for "cumulative exposure to respirable elemental carbon (REC) in µg/m^3-years".

MATERIAL

Vermeulen et al. 2014: input data for primary analysis

The meta-regression analysis of Vermeulen et al. [1] included three epidemiological US studies (Steenland et al. 1998 [2], Garshick et al. 2012 [3], Silverman et al. 2012 [4]) that estimated the association between occupational DEE exposure, measured as cumulative REC exposure in µg/m^3-years (dose), and lung cancer mortality.

Steenland et al. [2] is a nested case–control study on workers in the US trucking industry (994 lung cancer deaths and 1085 controls). The dose values were lagged 5 years when calculating the odds ratios (OR). Lagging is an evaluation technique which discards the exposure data of the last years (in this case the last 5 years) to take cancer latency phenomena into account [9]. This is the oldest study in the analysis, as it is based on the data from the case–control study by Steenland et al. [10]: "all cases and controls had died in 1982–1983." The measurements for elemental carbon by Zaebst et al. [11], which were not collected until 1990, were used as the basis for exposure estimation for the period from 1949 to 1983. Accordingly, the measurements were taken approx. 8 years after the death of the persons in the study, and therefore even later after the end of their exposure phase. Steenland et al. [2] attempted to extrapolate these data back "dependent on very broad assumptions". The authors evaluated this key limitation of their research as follows: "Our results should be regarded with appropriate caution because our exposure estimates are based on broad assumptions rather than actual measurements" and they noted the following in the abstract: "Our results depend on estimates about unknown past exposures, and should be viewed as exploratory."

Garshick et al. [3] is a cohort study independent of this on the US trucking industry (31,135 male employees, 779 lung cancer deaths). Date of death and cause-specific mortality was obtained from 1985 through 2000. Historical trends in ambient terminal REC were estimated based on historical trends in the coefficient of haze available for 1971 through 2000, a measurement of particulate matter based on optical density, assumed to be predictive of ambient REC. Vermeulen et al. [1] used this cohort data after excluding mechanics, and also used dose values lagged by 5 years. Those risk estimates (hazard ratios) from the Garshick study were used which were adjusted for duration of exposure.

Silverman et al. [4] is the case–control study of the US DEMS (DEMS = Diesel Exhaust in Miners Study). The underlying cohort [12] totalled 12,315 miners from 8 non-metal mining operations and included both surface and underground workers (no ore or coal mining). The case–control study included 198 lung cancer deaths and 562 controls based on mortality follow-up through December 31, 1997. Unlike the other two studies, the authors used a lag of 15 years to calculate the odds ratios. The exposures to REC were estimated in a complicated manner based on measurements of carbon monoxide (CO) and REC made in 1998–2000 and estimates of diesel equipment horsepower used through 1997 and mine ventilation.

The supplement to Vermeulen et al. [1] contains most of the input data to the Vermeulen meta-analyses (mean dose estimate for each dose category of the studies incorporated and the corresponding relative risk RR with a 95 % confidence interval). The data were extracted and transferred to a Stata file. Gaps were filled, where the original individual study publications contained the missing information.

Table 1 shows the input data as incorporated in the primary analysis by Vermeulen et al. [1]. The three studies included [2–4] reported cumulative exposure to DEE as respirable elemental carbon (dose) with relative risks and the corresponding 95 % confidence intervals. While the dose is reported using different categorizations, all three papers use μg/m^3-years as the dose unit. Table 1 lists ORs and HRs uniformly as relative risks.

Dose refers to the cumulative exposure to DEE in μg/m^3-years. For every category (Reference, Cat 1, Cat 2, Cat 3 and Cat 4), "averages" and the lower and upper limits of the dose are specified. Estimates of the relative risk RR with 95 % confidence interval (95 % CI) per category are given. Study size and number of lung cancer deaths are reported

[a]Garshick et al. [3] do not include mechanics as an employee group, and the risk estimates are adjusted for duration of exposure. For Garshick et al. [3], the number of person years is stated instead of the number of

persons, as it is a cohort study. Steenland et al. [2] specify neither the number of persons nor the number of cases per exposure category

Table 1. Input data on the primary analysis in Vermeulen et al. [1]

Study	lag/a	Exposure category	Average dose	Lower dose	Upper dose	RR	95 % CI		Number of persons	Number of cases
							Lower	Upper		
Steenland et al. 1998	5	Reference	0.0	0.0	0.0	1.00	1.00	1.00		
	5	Cat 1	84.5	0.0	<169.0	1.08	0.72	1.63		
	5	Cat 2	231.0	169.0	257.0	1.10	0.74	1.65		
	5	Cat 3	294.0	257.0	331.0	1.36	0.90	2.04		
	5	Cat 4	551.7	≥331.0		1.64	1.09	2.49		
Garshick et al. 2012[a]	5	Reference	15.5	0.0	<30.9	1.00	1.00	1.00	105513	122
	5	Cat 1	51.3	30.9	71.7	1.31	1.01	1.71	104909	191
	5	Cat 2	111.0	71.7	150.3	1.38	1.02	1.87	102496	202
	5	Cat 3	250.5	≥150.3		1.48	1.05	2.10	87397	226
Silverman et al. 2012	15	Reference	1.5	0.0	<3.0	1.00	1.00	1.00	207	49
	15	Cat 1	37.5	3.0	72.0	0.74	0.40	1.38	278	50
	15	Cat 2	204.0	72.0	536.0	1.54	0.74	3.20	206	49
	15	Cat 3	1036.0	≥536.0		2.83	1.28	6.26	173	50

For a more detailed description of the studies incorporated, we refer to Vermeulen et al. [1] and the original publications.

Corrected estimates: Garshick et al. 2012 (modified)

All analyses in Vermeulen et al. [1] applied the risk estimates (hazard ratios) from the Garshick study that were additionally adjusted for duration of exposure. An important aspect in this context is evident from the Letter to the Editor by Morfeld [13] (including the authors' answer) on adjustment errors in the coefficients used by Vermeulen et al. [1]. Morfeld criticised that the cumulative exposure was adjusted for duration of exposure, it already contains per definition. Thus, the risk coefficient does not estimate the effect of cumulative exposure, but that of a concentration (although this is not an optimal approach to estimating the concentration effect). The authors responded to the criticism as follows: "Morfeld suggests that adjusting cumulative exposure by duration of employment time reduces cumulative exposure to an estimate of long-term average concentration. We agree that if exposure in our workers was relatively constant, cumulative exposure would be the simple product of duration and average exposure. However, exposure varies considerably over time and between and within jobs." We emphasize that this note does not justify the procedure, as the data was evaluated using time-dependent methods in the Cox analyses performed. So the following applies for every point in time and in every person: cumulative exposure = duration of exposure × average concentration. Adjusting for duration of exposure changes the analysis such that time-dependent average concentration is analysed, not the cumulative exposure. The authors justify the exposure duration adjustment they made in spite of this by claiming that it controls for the healthy worker survivor effect. However, other methods are required for this [9].

Neither of the other studies [2, 4] made adjustments of this type to the cumulative exposure. Another aspect of the Garshick study, which also affects the model specification, is covered in the discussion section.

Garshick et al. [3] offered coefficients of cumulative exposure without adjustment for duration of exposure. In order to give the coefficients their usual meaning, and to facilitate comparison with the other two studies, these coefficients were also included in this analysis. The data are shown in Table 2 and will be referred to as Garshick et al. [3] (modified) below.

Table 2. Results from Garshick et al. [3] without adjustment for duration of exposure, referred to as Garshick et al. [3] (modified)

Exposure category	Average dose	Lower dose	Upper dose	RR	95 % CI		Person years	Number of cases
					Lower	Upper		
Reference	15.5	0.0	<30.9	1.00	1.00	1.00	105513	122
Cat 1	51.3	30.9	71.7	1.18	0.92	1.52	104909	191
Cat 2	111.0	71.7	150.3	1.17	0.88	1.55	102496	202
Cat 3	250.5	≥150.3		1.19	0.86	1.63	87397	226

Dose refers to the cumulative exposure to DEE in μg/m^3-years. For every category (Reference, Cat 1, Cat 2, Cat 3), averages and the lower and upper limits of the dose are specified. Odds ratio OR with 95 % confidence intervals (95 % CI) and the number of person years and the number of observed lung cancer deaths per dose category are reported

The risk estimates are not only lower than in Table 1, but also do not exhibit a positive trend with increasing exposure levels.

Vermeulen et al. [1] also performed sensitivity analyses with data deviating from Garshick et al. [3] as reported in Table 1. However, these variations only involved the lag of the exposure variables and the inclusion/exclusion of the data from mechanics. By contrast, the authors do not take the problem of the cumulative exposure coefficients incorrectly adjusted for duration of exposure into consideration.

Re-analysis of the DEMS case/control study: Crump et al. 2015

Crump et al. [14] re-analysed the DEMS case–control study by Silverman et al. [4] and largely managed to reproduce its results: "We were able to replicate the findings reported by Silverman et al. (18) when we used the same analytical methods. This gave us confidence that we were using the same basic data set as Silverman et al.". Crump et al. investigated the influence of covariables which Silverman et al. [4] did not include in their final models. The radon exposure underground proved to be a main confounder, a result which did not match the statements by Silverman et al. [4]. On the cumulative exposure to radon, Silverman et al. [4] wrote "estimated cumulative exposure to radon ... were evaluated but not included in the final models because they had little or no impact on odds ratios (i.e., inclusion of these factors in the final models changed point estimates for diesel exposure by ≤10 %)". Crump et al. [14] noted the following on this: "However, when we reproduced the Silverman et al. analysis, we could not verify this statement." As a result, the present sensitivity meta-analysis only uses the estimates by Crump et al. [14] after an additional adjustment for cumulative radon exposure.

Crump et al. [14] also developed six new DEE exposure metrics, as an alternative to the estimates used in Attfield et al. [12] and Silverman et al. [4]: "We proceeded to apply six alternative REC metrics, five of which depended, as did the DEMS metrics, on extrapolations involving assumed relationships between CO and REC. A sixth REC metric, REC6, was used that did not involve any assumptions concerning the relationship between CO and REC, and was based on Adj_HP [adjusted horse power] and ventilation rates for each of the mines. Of the several REC metrics, we view REC6 as having some superior qualities because it avoids using the highly uncertain assumptions concerning the relationship between CO and REC." Therefore, we use REC6 as a primary alternative to the Silverman exposure data.

In their supplementary evaluations for the exposure estimates, Crump and van Landingham [15] determined $\beta = 0.3$ as the best estimate in the $REC \sim CO^{\beta}$ conversion model, as compared with the value of $\beta = 0.58$

according to Stewart et al. [16] or $\beta = 1.0$, which Silverman et al. [4] assumed in their analyses. The alternative exposure metric REC4 of Crump et al. [14] contains this best estimate, $\beta = 0.3$. REC4 is also directly based on the work by Crump and van Landingham [15] to estimate an alternative exposure, and takes several other aspects into consideration (see the detailed description of REC1 to REC4 in Crump et al. [14]). As a result, we also used the exposure metric REC4 as a second variation according to Crump et al. [14]. Table 3 gives an overview of the results of the re-analysis with REC4 and REC6 according to Crump et al. [14]. For the sake of clarity we emphasize that we only used results based on REC4 and REC6 after adjustment for radon exposure.

Table 3. Re-analysis of Silverman et al. [4] with adjustment for radon exposure (Crump et al. [14], Table 3)

Exposure category	Average dose	Lower dose	Upper dose	OR	95 % CI		Number of persons	Number of cases
					Lower	Upper		
Crump et al. [14], REC4								
Reference	0.71	0	<4.9	1	1	1	217	49
Cat 1	26.65	4.9	<70.4	0.80	0.41	1.55	266	50
Cat 2	243.43	70.4	<498.4	1.67	0.73	3.81	192	49
Cat 3	1522.10	-	≥498.4	1.50	0.54	4.17	189	50
Crump et al. [14], REC6								
Reference	0.61	0	<2.8	1	1	1	230	49
Cat 1	20.47	2.8	<50.6	1.07	0.57	2.00	247	50
Cat 2	158.27	50.6	<388.0	1.35	0.62	2.94	206	49
Cat 3	1156.89	-	≥388.0	1.43	0.52	3.94	181	50

The results are based on the exposure metrics REC4 and REC6 with a lag of 15 years (see text). Dose refers to the cumulative exposure to DEE in $\mu g/m^3$-years. For every category (Reference, Cat 1, Cat 2, Cat 3), averages and the lower and upper limits of the dose are specified. Odds ratio OR with 95 % confidence intervals (95 % CI) and the number of persons observed and the number of lung cancer deaths per dose category are reported

Re-categorised data: Möhner et al. 2013 (adapted)

Vermeulen et al. [1] omitted the German potash miner study of Möhner et al. [17] from their analysis, as they alleged the reference category defined was too high. Möhner et al. [17] was only used in two sensitivity analyses, however it was incorporated by Vermeulen et al. [1] either leaving the original RR estimates or correcting the risk estimates ad-hoc. Both approaches are less than ideal, which is why Vermeulen et al. [1] did not present either of the evaluations as part of their primary analysis, merely reporting on the results in the online appendix.

The German Potash miner study included 5,819 miners followed in mortality from 1970 through 2001. In 1991, exposure measurements of the concentration of total and elemental carbon (to a lesser extent) in the respirable dust fraction by coulometric analysis were undertaken. Because the mining technology and the mining equipment remained fairly stable since 1969, measurements from 1991 have been used for designing a job-exposure-matrix with three main job categories: production, maintenance, and workshop. Elemental carbon was the largest component of total carbon with a proportion of weight of about 63 %. Moreover, the two measures were highly correlated ($r = 0.89$). Cumulative exposure was determined as REC in µg/m^3-years.

In order to incorporate the German potash miner study as informatively as possible in this project, the data had to be used at a different scale to that published. We contacted the original authors for the information required to do so (the publication does not contain all relevant categorizations of exposure, and other detailed information was missing). On request, Dr. Möhner provided additional results from the Möhner et al. [17] study by e-mail on 03/07/2014. This included case numbers and odds ratio estimates with 95 % confidence intervals for a modified categorisation, referred to as “Möhner et al. [17] (adapted)” below.

The breakdown was adapted so that the reference category only consists of 5 cases, with the result that the exposure level of the reference category is far lower than in the original analysis (a criticism in Vermeulen et al. [1]). The other three categories were chosen in such a way that the case distribution is roughly equal. Table 4 shows the results of this re-categorisation of Möhner et al. [17].

Dose refers to the cumulative exposure to DEE in µg/m^3-years. For every category (Reference, Cat 1, Cat 2, Cat 3), averages and the lower and upper limits of the dose are specified. Odds ratio OR with 95 % confidence intervals (95 % CI) and the number of persons observed and the number of lung cancer deaths per dose category are reported

Table 4. Additional results for the study by Möhner et al. [17] with changed (adapted) categorisation, referred to as Möhner et al. [17] (adapted)

Exposure category	Average dose	Lower dose	Upper dose	OR	95 % CI		Number of persons	Number of cases
					Lower	Upper		
Reference	263	0	≤380	1	1	1	29	5
Cat 1	728	380	1024	0.99	0.323	3.031	126	19
Cat 2	1386	1024	1840	1.69	0.53	5.403	126	25
Cat 3	2538	>1840		1.2	0.363	3.971	127	19

METHODS

Reproduction of the results and variation of modelling

The complex meta-analysis of Vermeulen et al. [1] of the categorised results of the individual studies was replicated. The exposure categorisation, which varied with the studies, and the potential differences in the design and size of the studies were a challenge. In addition to this, the results for the exposure categories are nested within the studies, creating a two-level structure (1st level: exposure groups, 2nd level: studies). In order to analyse this complex data situation appropriately, meta-regression analyses were performed on the results of the individual studies, both with fixed effects [18] and with random effects [19–21], to determine a shared exposure-response curve (see also [22]).

The analyses must be weighted by variances of the RR estimates per exposure category. This weighting is required as only aggregated data (i.e., no individual data on persons) are available in the evaluations, and the data points for the exposure categories differ substantially in a joint evaluation due to the varying number of cases within and between the studies. Different approaches were trialled for this:

- Linear regression with fixed effects for log RR with weights proportional to the inverse of the respective variance
    - Without adjustment by study
    - With adjustment by study. A global F-test on the heterogeneity of the risk levels (intercepts, offsets) between the studies was performed.
- Mixed linear regression for log RR with a random intercept incorporating the differences between the studies

    - With weights at the first level (exposure categories level) proportional to the inverse of the respective variance and with the totals of these weights as study weights at the second level (study level)
    - In a second evaluation approach, the weights of the first level were scaled effectively to the weights of the second level.

- Mixed linear regression for log RR with a random intercept and a random dose coefficient (slope) incorporating the differences between the studies
    - With weights at the first level (exposure categories level) proportional to the inverse of the respective variance and with the totals of these weights as study weights at the second level (study level)
    - In a second evaluation approach, the weights of the first level were scaled effectively to the weights of the second level.

The analyses do not take the reference points into account, as their weights are infinite, i.e., not a real number. That corresponds to the procedure in Vermeulen et al. [1]: Fig. 1 only contains 10 observation points, not 13.

Models with fixed effects are usually evaluated statistically via the Student's t-distribution [23]. As only a few data points are incorporated, which reflect groups rather than individuals (aggregated data), this tends to lead to over-estimations of P values and confidence interval widths. However, precision weightings only include the relative weighting differences between the data points in the analyses and, thus, does not solve this problem. Therefore, the models with fixed effects are evaluated based on the standard normal distribution as an alternative [23]. This tends to lead to P values and confidence intervals which are too narrow, in particular as the correlations between the groups within the studies are not taken into consideration. We note that all mixed regressions are always evaluated via the standard normal distribution.

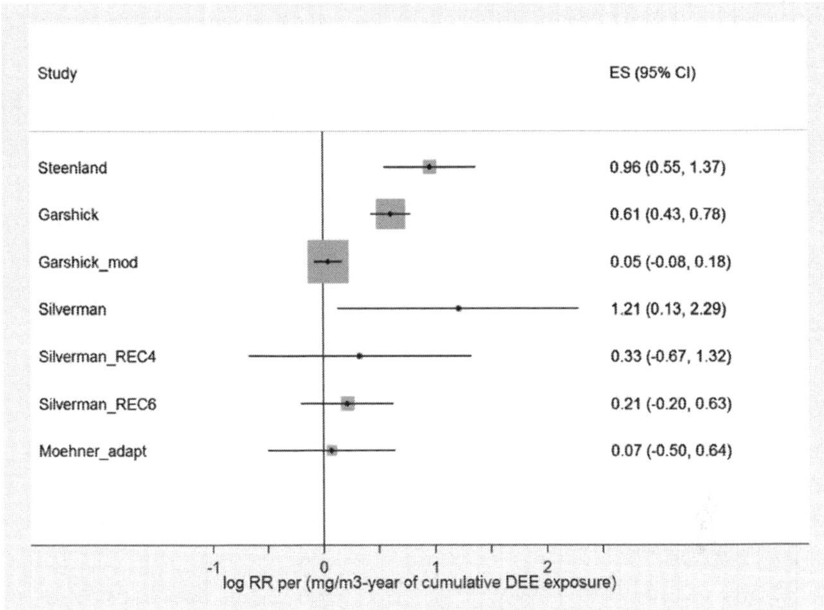

Figure 1. Analysis of the individual studies. Linear regression with fixed effects for log RR with weights proportional to the inverse of the respective variance (represented by box size). Coefficients (effect size, ES) of the cumulative exposure (Dose) to DEE in mg/m^3-years with a 95 % confidence interval for the coefficient, calculated using the standard normal distribution. Steenland et al. [2] and Silverman et al. [4] were evaluated as used by Vermeulen et al. [1] as well as in accordance with the re-analysis by Crump et al. [14] with adjustment for the radon exposure (REC4, REC6, cf. Table 3), as was Garshick et al. [3] as used by Vermeulen et al. [1], and Garshick et al. [3] (modified) and the data based on Möhner et al. [17] (adapted)

However, the results reported in a study for the various exposure categories are not independent of one another, but correlate with one another, as they refer to a common reference category within this study. As a result, in addition to the evaluations described above, meta-regression methods including the correlations of the study results within the studies were attempted [24–26]. However, additional input is required to use this Greenland/Longnecker method: person years and case numbers or person numbers and case numbers per exposure category from the studies must be available. These data are not included in the supplement to Vermeulen et al. [1], and cannot be reconstructed in full from the original publications.

In their Methods section, Vermeulen et al. [1] mentioned almost all of these methods, but it remains unclear whether they always used the Greenland/Longnecker method, for instance, and which weighting structure was used in the mixed regressions. With different evaluation approaches, this research project aims to evaluate the extent to which and the method with which the results published by Vermeulen et al. [1] can be reproduced.

Vermeulen et al. [1] also used spline models, which permit a more flexible curve shape than log-linear models. However, these spline regressions did not result in a deviating curve for the estimated exposure-response relationship: "The linear model (Fig. 1) and the spline meta-regression model (data not shown) fit the data well, with virtually equivalent curves." That is why we did not use any spline functions or similar methods, instead following the main approach of Vermeulen et al., also to keep the number of parameters to be estimated as low as possible.

As the main modelling approach, a precision-weighted regression analysis with fixed effects and simultaneous adjustment for the individual studies was pursued, as it is more stable and easier to interpret. This approach also has other advantages over a regression with random effects (see the justifications in Allison [18], pp. 2, 3, and Cameron and Trivedi [19], p. 700). We examined whether the other approaches outlined above offer relevantly different results. If that is not the case, this method is used as the leading analytical strategy.

All analyses were calculated using Stata 12 [27].

Influence of the input data selected

One important aspect of data selection can be derived from the Letter to the Editor by Morfeld [13] regarding the Garshick et al. [3] paper, stating that the coefficients used by Vermeulen et al. [1] are incorrectly adjusted. Therefore, another analysis was performed to repeat the meta-analyses described above with the corrected coefficients per Garshick et al. [3] (modified) (cf. Table 2).

The results of the DEMS re-analysis were presented in an HEI (Health Effects Institute) webinar [28,29] and published in detail afterwards

[14, 30]. Crump et al. [14] contained revised OR estimates on Silverman et al. [4]; important alternative estimates from this paper are presented in Table 3. The "REC6" findings from this paper are incorporated in the meta-analysis after adjustment for the radon exposure instead of the data from Silverman et al. [4] (cf. Table 1). Variation in the results when using the "REC4" findings after adjustment for radon exposure was also investigated.

The German potash miner study [17] was also incorporated as part of a sensitivity analysis. We used the risk estimates from "Möhner et al. [17] (adapted)" (cf. Table 4).

As the paper by Steenland et al. [2] has significant limitations (see the "Material" section), meta-analyses were also performed without including this study.

Non-linearity: threshold search

We examined the data for non-linearities in the exposure-risk relationship. In particular, a systematic search was performed for cumulative exposure thresholds [31–33]. To do so, the algorithm presented in detail in Morfeld et al. [34] to determine a no-adverse effect level in the cumulative exposure was combined with the meta-regression methods. This analysis was restricted to the main modelling approach (precision-weighted regression analysis with fixed effects and simultaneous adjustment for individual studies).

10 $\mu g/m^3$-years was selected as the increment for threshold exploration (range: 0 $\mu g/m^3$-years to 500 $\mu g/m^3$-years). Accordingly, 51 models were calculated and compared per threshold search. The threshold analysis was made using special programs in Stata 12 [27].

RESULTS

Reproduction of the results in Vermeulen et al. [*1*] and extended analyses

In Table 1 (primary analysis), Vermeulen et al. [1] reported on the three studies by Steenland et al. [2], Garshick et al. [3] and Silverman et al. [4]. The authors stated individual findings (risk coefficients calculated) and the result of the meta-analysis of the risk coefficients. We compare these findings with the results of the re-analysis below.

Analyses of the individual studies

All individual analyses were recalculated with fixed effects linear regression evaluating log RR (for methodological reasons, this analysis is identical to the mixed regression analysis if confined to a single study). Tables 5, 6, 7, 8 show the results for Steenland et al. [2], Garshick et al. [3], Silverman et al. [4] and Möhner et al. [17].

Table 5. Steenland et al. [2]: Linear regression with fixed effects for log RR with weights proportional to the inverse of the respective variance

Log RR	Coef.	Std. Err.	t	*P*	95 % CI	
Dose	0.00096	0.00021	4.55	0.045 (<0.001)	0.00005 (0.00055	0.00187 0.00137)
Constant	−0.031	0.070	−0.45	0.70	−0.33	0.27

Coefficient (Coef.) of the cumulative exposure (dose) to DEE in μg/m³-years and absolute term (constant) with standard deviation (Std.Err.); t-test statistic for the coefficient with the corresponding P-value and a 95 % confidence interval for the coefficient, 95 % CI. The figures in parentheses are the results when using the standard normal distribution

Table 6. Garshick et al. [3]: Linear regression with fixed effects for log RR with weights proportional to the inverse of the respective variance

log RR	Coef.	Std. Err.	t	*P*	95 % CI	
Dose	0.00061	0.000091	6.6	0.095 (<0.001)	−0.00055 (0.00042	0.00177 0.00078)
Constant	0.244	0.013	18.6	0.034	0.078	0.411
Garshick (modified)						
Dose	0.00005	0.00007	0.75	0.59 (0.45)	−0.00077 (−0.00008	0.00087 0.00017)
Constant	0.159	0.009	16.87	0.038	0.039	0.279

Coefficient (Coef.) of the cumulative exposure (Dose) to DEE in μg/m³-years and absolute term (Constant) with standard deviation (Std.Err.); t-test statistic for the coefficient with the corresponding P-value and a 95 % confidence interval for the coefficient, 95 % CI. The figures in parentheses are the results when using the standard normal distribution. Garshick et al. [3] as used by Vermeulen et al. [1], and Garshick et al. [3] (modified) were evaluated

Table 7. Silverman et al. [4]: Linear regression with fixed effects for log RR with weights proportional to the inverse of the respective variance

log RR	Coef.	Std. Err.	t	*P*	95 % CI	
Dose	0.00121	0.00055	2.20	0.27 (0.028)	−0.00579 (0.00013	0.00821 0.00229)
Constant	−0.148	0.299	−0.49	0.71	−3.95	3.65
Crump et al. [14], REC4						
Dose	0.00033	0.00051	0.64	0.63 (0.53)	−0.00615 (−0.00067	0.00680 0.00132)
Constant	0.00541	0.358	0.02	0.99	−4.54	4.55
Crump et al. [14], REC6						
Dose	0.00021	0.00021	1.01	0.50 (0.31)	−0.00247 (−0.00020	0.00289 0.00063)
Constant	0.137	0.108	1.27	0.42	−1.23	1.51

Coefficient (Coef.) of the cumulative exposure (Dose) to DEE in µg/m^3-years and absolute term (Constant) with standard deviation (Std.Err.); t-test statistic for the coefficient with the corresponding P-value and a 95 % confidence interval for the coefficient, 95 % CI. The figures in parentheses are the results when using the standard normal distribution. Silverman et al. [4] was evaluated as used by Vermeulen et al. [1], and in accordance with the re-analysis by Crump et al. [14] with adjustment for radon exposure (REC4, REC6, cf. Table 3)

Table 8. Möhner et al. [17] (adapted): Linear regression with fixed effects for log RR with weights proportional to the inverse of the respective variance

log RR	Coef.	Std. Err.	t	*P*	95 % CI	
Dose	0.00007	0.00029	0.24	0.85 (0.81)	−0.00364 (−0.00050	0.00378 0.00064)
Constant	0.119	0.492	0.24	0.85	−6.13	6.37

Coefficient (Coef.) of the cumulative exposure (Dose) to DEE in µg/m^3-years and absolute term (Constant) with standard deviation (Std.Err.); t-test statistic for the coefficient with the corresponding P-value and a 95 % confidence interval for the coefficient, 95 % CI. The figures in parentheses are the results when using the standard normal distribution

Table 1 in Vermeulen et al. [1] reports the following coefficient estimates for Steenland et al. [2]: dose = 0.00096 (95 % CI: 0.00033, 0.00159), constant = −0.032. The agreement in the coefficients is very good. The confidence interval is consistent when both calculation methods are taken into account (Student's t-distribution, normal distribution).

Table 1 in Vermeulen et al. [1] reported the following coefficient estimates for Garshick et al. [3]: dose = 0.00061 (95 % CI: −0.00088,

0.00210), constant = 0.24. This coefficient estimate is also very similar, and the interval estimate is a good match. The estimates differ clearly when we evaluate Garshick et al. [3] (modified): The dose coefficient is lower by more than an order of magnitude, and far from being significant ($P > 0.4$). A significant deviation of the constant (intercept) from the normal value zero is apparent. This normal value zero means a baseline level of the relative risk of 1, as is to be assumed for a DEE exposure of 0 µg/m³-years.

Table 1 in Vermeulen et al. [1] reported the following coefficient estimates for Silverman et al. [4]: dose = 0.00120 (95 % CI: 0.00053, 0.00187), constant = –0.18. The coefficient estimates here also agree well (there is probably a typing error in Vermeulen et al. [1]: –0.18 instead of –0.148). The interval estimates are compatible too.

By contrast, the dose coefficients calculated using the Crump variations are far lower than the original values: They are only 27 % (REC4) or 18 % (REC6) of the Silverman coefficient. In these calculations, the t-test confidence intervals are largely symmetrical around zero, so that neither model reveals a trend. Even using the standard normal distribution, there is no indication of a significant influence of the DEE exposure on lung cancer mortality when REC4 or REC6 are used as the exposure metrics. We like to emphasize that REC4 and REC6 are both adjusted for radon exposure as a potential confounder.

Table 8 shows the results obtained by evaluating the study by Möhner et al. [17] with changed (adapted) exposure categorisation. The German study does not indicate an association between DEE exposure and lung cancer mortality.

Figure 1 gives an overview of the results of the regressions with fixed effects, assuming a normal distribution for statistical evaluation of all individual studies. Models with fixed effects are presented to prevent underestimation of the coefficients. We selected the normal distribution to rule out overestimation of the p-values (i.e., the confidence intervals presented are definitely not too wide). Deviating from Vermeulen et al. [1], we selected mg/m³-years = 1000 × µg/m³-years as the exposure unit, to keep the overview clearer. Each of the three data sets incorporated in the Vermeulen analysis [2–4] results in significantly elevated risk estimates, whereby Garshick et al. [3] paper has the greatest influence. When we analysed the results from Garshick et al. [3] without adjusting for exposure duration, i.e., Garshick et al. [3] (modified), the risk estimate drops considerably and is not significant.

The Crump modifications (REC4, REC6) of the research in Silverman et al. [4] also result in considerably lower risk estimates than in the original Silverman et al. paper, and are not statistically significant. The study by Möhner et al. [17] does not give any indication of an association between DEE exposure and lung cancer mortality.

Joint analysis of the studies

According to Table 1 in Vermeulen et al. [1], the following values apply for "all studies combined" (i.e., [2–4]) as the coefficients of cumulative exposure (dose) to DEE in µg/m^3-years and the absolute term (constant): dose = 0.00098 (95 % CI: 0.00055, 0.00141) and constant = 0.088. Table 9 reports on the re-analysis of the three studies using a log-linear regression with fixed effects, without adjusting for the studies.

Table 9. Linear regression with fixed effects for log RR with weighs proportional to the inverse of the respective variance, without adjustment for studies

log RR	Coef.	Std. Err.	t	P	95 % CI	
Dose	0.00076	0.00026	2.97	0.018 (0.003)	0.00017 (0.00026	0.00135 0.00126)
Constant	0.130	0.072	1.80	0.11	−0.0365	0.297
Crump et al. [14], REC4						
Dose	0.00033	0.00022	1.46	0.18 (0.15)	−0.00019 (−0.00011	0.00084 0.00076)
Constant	0.208	0.067	3.10	0.015	0.053	0.336
Crump et al. [14], REC6						
Dose	0.00034	0.00020	1.74	0.12 (0.083)	−0.00011 (−0.00004	0.00080 0.00073)
Constant	0.211	0.054	3.90	0.005	0.860	0.335

Coefficient (Coef.) of the cumulative exposure (dose) to DEE in µg/m^3-years and absolute term (Constant) with standard deviation (Std.Err.); t-test statistic for the coefficient with the corresponding P-value and a 95 % confidence interval for the coefficient, 95 % CI. The figures in parentheses are the results when using the standard normal distribution. Silverman et al. [4] was incorporated as used by Vermeulen et al. [1], and in accordance with the re-analysis by Crump et al. [14] with adjustment for radon exposure (REC4, REC6, cf. Table 3)

Table 9 shows relevant deviations from the results in Vermeulen et al. [1]. While the exposure-response association is statistically significant, the dose coefficient estimated without adjustment for the studies is approx. 20 % lower than that published by Vermeulen and colleagues (0.00076/0.00098 = 0.77). If the versions of the results according to Crump et al. [14] are incorporated in the evaluation, the meta-analysis does not indicate a significant association between DEE exposure and lung cancer mortality.

Table 10 also adjusts the coefficient estimates by studies.

Table 10. Linear regression with fixed effects for log RR with weights proportional to the inverse of the respective variance, with adjustment for studies

log RR	Coef.	Std. Err.	t	*P*	95 % CI	
Dose	0.00106	0.00020	5.35	0.002 (<0.001)	0.00057 (0.00067	0.00154 0.00144)
ΔConst1	0.249	0.080	3.11	0.021	0.052	0.445
ΔConst2	−0.035	0.126	−0.28	0.79	−0.343	0.272
Constant	−0.059	0.079	−0.74	0.49	−0.252	0.135
Crump et al. [14], REC4						
Dose	0.00053	0.00021	2.49	0.047 (0.013)	0.00001 (0.00011	0.00105 0.00094)
ΔConst1	0.163	0.095	1.71	0.14	−0.070	0.396
ΔConst2	−0.167	0.169	−0.98	0.36	−0.581	0.248
Constant	0.091	0.091	1.01	0.35	−0.130	0.313
Crump et al. [14], REC6						
Dose	0.00054	0.00017	3.12	0.021 (0.0018)	0.00012 (0.00020	0.00096 0.00087)
ΔConst1	0.164	0.069	2.40	0.054	−0.003	0.332
ΔConst2	0.041	0.113	−0.37	0.73	−0.319	0.236
Constant	0.088	0.068	1.30	0.24	−0.079	0.255

Coefficient (Coef.) of the cumulative exposure (Dose) to DEE in µg/m^3-years and absolute term (Constant) with standard deviation (Std.Err.); t-test statistic for the coefficient with the corresponding P-value and a 95 % confidence interval for the coefficient, 95 % CI. The figures in parentheses are the results when using the standard normal distribution. Silverman et al. [4] was incorporated as used by Vermeulen et al. [1], and in accordance with the re-analysis by Crump et al. [14] with adjustment for radon exposure (REC4, REC6, cf. Table 3)

Constant: Steenland et al. [2] (reference study)

ΔConst1: Garshick et al. [3] - Steenland et al. [2]

ΔConst2: Silverman et al. [4] - Steenland et al. [2]

Table 10 shows a good correlation for the dose coefficient and the corresponding confidence interval with the results reported in Vermeulen et al. [1] (dose = 0.00098, 95 %-CI: 0.00055, 0.00141; constant = 0.088). The recalculated coefficient exhibits minimal upward deviation. The two confidence interval calculations (Student's t-distribution, normal distribution) have largely matching results, so that the 10 data points are sufficient to permit a robust estimate, which also agrees to the results published by Vermeulen and his colleagues. The global F-test on the heterogeneity of the baseline risk levels between the studies, which was also calculated, results in $F(2, 6) = 5.5$, $P = 0.044$. Thus, the heterogeneity between the three studies is significant at a 5 % level. Heterogeneity refers to a systematic difference in the baseline risk of the three studies (i.e., setting exposure to zero). For a meaningful combination of the three studies, their base levels should match except for random deviations. However, the differences are statistically significant. The heterogeneity between the three studies results from a significant difference between Garshick et al. [3] and Steenland et al. [2], ΔConst1: $p = 0.021$. Accordingly, the Garshick study deviates significantly upwards in the baseline level of the risk from the other two studies, as was already indicated in the individual analysis (Table 6). Correcting the Garshick coefficients reduces the heterogeneity in the risk level considerably: $F(2, 6) = 0.50$, $P = 0.63$. This is not apparent from the individual analysis.

If we replace the original data per Silverman et al. [4] with the results from Crump et al. [14] (REC4 or REC6, both adjusted for radon exposure), the dose coefficients in the meta-regression are halved.

Table 11 shows results when a mixed linear regression for the three studies is fitted.

Table 11 results in a very good agreement to the finding published in Vermeulen et al. [1] (dose = 0.00098, 95 % CI: 0.00055, 0.00141; constant = 0.088). If the dose coefficient is also estimated as a random effect, the results do not change. The findings deviate slightly from Vermeulen et al. [1] without an effective scaling of the weights (result: dose coefficient = 0.00093, constant =0.096). If Garshick et al. [3] (modified) is included in the meta-analysis, the estimated meta-risk coefficient and the corresponding significance level decreases (test statistic changes from $Z = 6.74$ to $Z = 4.58$), however the positive dose–response association is always significant (see Table 11).

Table 11. Mixed linear regression for log RR with a random intercept with weights at the first level (exposure categories level) proportional to the inverse of the respective variance and with the totals of these weights as study weights at the second level (study level)

log RR	Coef.	Robust Std. Err.	Z	P	95 % CI	
Dose	0.00097	0.00014	6.74	<0.001	0.00069	0.00125
Constant	0.084	0.111	0.75	0.45	−0.134	0.300
Garshick (modified)						
Dose	0.00087	0.00019	4.58	<0.001	0.00050	0.00124
Constant	0.024	0.062	0.39	0.70	−0.097	0.145
Crump et al. [14], REC4						
Dose	0.00045	0.00021	2.10	0.035	0.00003	0.00086
Constant	0.178	0.094	1.90	0.057	−0.005	0.362
Crump et al. [14], REC4 and Garshick (modified)						
Dose	0.00048	0.00023	2.10	0.036	0.00003	0.00092
Constant	0.097	0.040	2.39	0.017	0.018	0.176
Crump et al. [14], REC6						
Dose	0.00053	0.00026	2.05	0.040	0.00002	0.00103
Constant	0.172	0.101	1.71	0.088	−0.025	0.369
Crump et al. [14], REC6 and Garshick (modified)						
Dose	0.00052	0.00025	2.09	0.036	0.00003	0.00101
Constant	0.095	0.051	1.87	0.061	−0.044	0.194

The weights of the first level are scaled effectively to the weights of the second level. Coefficient (Coef.) of the cumulative exposure (Dose) to DEE in µg/m^3-years and absolute term (Constant) with standard deviation (Std.Err.); Z-test statistic for the coefficient with the corresponding P-value and a 95 % confidence interval for the coefficient, 95 % CI. Silverman et al. [4] was incorporated as used by Vermeulen et al. [1], and in accordance to the re-analysis by Crump et al. [14] with adjustment for the radon exposure (REC4, REC6, cf. Table 3). Garshick et al. [3] as used by Vermeulen et al. [1], and Garshick et al. [3] (modified) were incorporated

As in the models with fixed effects (Table 10), the dose coefficients in the meta-regression are halved in the mixed regression (Table 11) if the original data in accordance with Silverman et al. [4] is replaced with the findings from Crump et al. [14] (REC4 or REC6, both adjusted for radon exposure).

Crump [39] also recalculated some of the Vermeulen et al. [1] results using a mixed model, but did not see any way to obtain the data required to use the Greenland/Longnecker method (cf. our explanations in the Method section "Reproduction of the results and variation of modelling"). Crump wrote: "I ... reran the analysis of Vermeulen et al. [1], except that I did not model the dependence among the ORs from the same study. (I did not have access to data needed to model that dependence.) My analysis yielded a regression parameter [0.88; 95 % confidence interval (CI): 0.65, 1.11] similar to that obtained by Vermeulen et al. [1] (0.98; 95 % CI: 0.55, 1.41)." Crump [29] obviously selected a different unit of DEE exposure, at 1000 µg/m^3-years. The dose coefficient (0.88) he reported deviates more than the result of the recalculation performed in this paper (0.97), which corresponds very well with the Vermeulen result (0.98). It is not clear whether Crump scaled the weights, as even the recalculation in this report is slightly lower (0.93) without scaling, though not as pronounced as in Crump [29].

Variation of modelling

As Steenland et al. [2] did not provide details on the person years and case numbers per exposure category, this paper cannot be included in evaluations with the Greenland/Longnecker method. However, Garshick et al. [3], Silverman et al. [4] and Möhner et al. [17] (adapted) can be analysed jointly. Table 12 shows the results.

Table 12 shows that the linear regressions with fixed effects for log RR with weights proportional to the inverse of the respective variance and with adjustment for studies result in slightly more pronounced effect coefficients and somewhat lower P-values than models with random effects (that qualitatively matches the findings presented in the Results section: "Joint analysis of the studies"). However, if the dose coefficient is also calculated as a random effect, a similar value results (coefficient = 0.00059, P = 0.17) for the model with fixed effects. By contrast, the analyses with the Greenland/Longnecker method, recommended by methodologists and taking internal correlations into account, resulted in a lower coefficient than in the model with fixed effects, with a lower P-value = 0.035 (significant). In this joint analysis of Garshick et al. [3] (modified coefficients), Silverman et al. [4] and Möhner et al. [17] (adapted) using the Greenland/Longnecker method, the dose coefficient is approximately a factor of 3 lower than the estimate in the primary analysis by Vermeulen et al. [1] (0.0032 vs. 0.00098).

Table 12. Joint analysis of Garshick et al. [3] (modified coefficients), Silverman et al. [4] and Möhner et al. [17] (adapted)

	Coef.	*P*	95 % CI	
Fixed effects, adjusted	0.00054	0.14 (0.083)	−0.00026 (−0.00007	0.00133 0.00115)
Random intercept, scaled	0.00024	0.36	−0.00027	0.00074
Greenland/Longnecker method	0.00032	0.035	0.00002	0.00062
Crump et al. [14], REC4				
Fixed effects, adjusted	0.00020	0.35 (0.30)	−0.00029 (−0.00018	0.00069 0.00058)
Random intercept, scaled	0.00010	0.11	−0.00002	0.00022
Greenland/Longnecker method	0.00015	0.31	−0.00014	0.00044
Crump et al. [14], REC6				
Fixed effects, adjusted	0.00013	0.33 (0.28)	−0.00018 (−0.00010	0.00042 0.00035)
Random intercept, scaled	0.000073	0.025	0.00001	0.00014
Greenland/Longnecker method	0.00012	0.44	−0.00019	0.00043

Comparison of the estimated effect for the cumulative exposure to log RR, calculated with various regression methods. Coefficient (Coef.) of the cumulative exposure (Dose) to DEE in µg/m^3-years, corresponding P-value and 95 % confidence interval for the coefficient, 95 % CI. The figures in parentheses are the results when using the standard normal distribution in the model with fixed effects. Silverman et al. [4] was incorporated as used by Vermeulen et al. [1], and in accordance with the re-analysis by Crump et al. [14] with adjustment for radon exposure (REC4, REC6, cf. Table 3)

The results reported by Vermeulen et al. [1] largely match the estimates in the models with random effects (cf. Results section on the "Joint analysis of the studies"). However, the Greenland/Longnecker method cannot be used in this case without access to further data. Vermeulen and his colleagues were obviously given access to these data from Steenland et al. [2], but did not disclose them in their paper, even in the supplement.

If we replace the original data per Silverman et al. [4] with the results from Crump et al. [14] (REC4 or REC6, after adjustment for radon exposure), the statements on the three analysis methods are qualitatively the same. Accordingly, the dose coefficients in the meta-regression are halved due to the data variation. In spite of the different models, the P-values are largely equivalent. One exception to this is regression with a random intercept for REC6, where the value is lower ($P = 0.25$). Like the deviating result of the Greenland/Longnecker regression on evaluation with the Silverman original data ($P = 0.035$), this indicates instabilities in the variance estimate.

In order to avoid discussions on a potential underestimation of the effect, all further analyses were made using adjusted regression models with fixed effects. In Tables 11 and 12, the dose coefficient estimates (fixed effects adjusted for studies and random effects) almost match, and are also almost identical with those in Vermeulen et al. [1]. A comparison of the interval estimates and P-values indicates a greater stability of the regression models with fixed effects. This justifies the decision to use adjusted regression models with fixed effects as a primary analysis method (main modelling approach) for this research project.

Influence of the input data selected

Regardless of the method selected, a simultaneous analysis of the three studies by Garshick et al. [3] (modified), Silverman et al. [4] and Möhner et al. [17] (adapted) results in a far lower effect coefficient in Table 12 than in the primary analysis by Vermeulen et al. [1]. Example: For the regression models with fixed effects and adjustment for studies, a reduction in the effect coefficient by approx. 50 % resulted, as the reproduced primary analysis in accordance with Vermeulen et al. [1] led to a coefficient of 0.0011 (per 1 $\mu g/m^3$-years), $P = 0.002$, and a 95 % CI of 0.00057 to 0.00154 (cf. Table 10).

An interesting finding to be noted is that a simultaneous analysis of the three studies by Garshick et al. [3] (modified), Silverman et al. 2012 and Möhner et al. [17] (adapted) only finds a significant dose–response relationship (Table 12) if the Greenland/Longnecker method is used. This could be due to instabilities in estimation in this complex procedure.

If the primary analysis in accordance with Vermeulen et al. [1] is performed with fixed effects and adjustment for studies and with modified Garshick coefficients, this results in a coefficient of 0.00098 (per 1 µg/m^3-year), $P = 0.006$, and a 95 % CI of 0.00041 to 0.00155, somewhat more pronounced but similar to the results of the regression calculation with random effects (cf. Table 11).

Thus, the modification of the Garshick coefficients only reduces the meta-coefficient by approx. 7 % and the statistical significance is maintained, although the P-value increases from 0.002 to 0.006. Correcting the Garshick coefficients reduces the heterogeneity in the base level of the risk considerably: $F(2, 6) = 0.50$, $P = 0.63$. Without correction we get: $F(2, 6) = 5.5$, $P = 0.044$.

Tables 10, 11, 12 are consistent in showing that replacing the original data from Silverman et al. [4] with the results from Crump et al. [14] (REC4 or REC6, both with adjustment for radon) reduces the dose coefficient by approximately half in the meta-regressions. If we analyse Steenland et al. [2], Crump et al. [14] and Garshick et al. [3] (modified) together, the dose–response relationship in adjusted regression models with fixed effects and evaluation with the standard normal distribution is significant (REC4: $P = 0.03$, REC6: $P = 0.01$), but clearly weaker than when analysing the original data used by Vermeulen et al. [1] ($P < 0.000001$). The lower confidence interval limits of the dose coefficient are correspondingly different: 0.00005, 0.00010 and 0.00067. They differ by a factor of at least 12 or 6.

Figure 2 gives an overview of the results of the meta-regressions with fixed effects for variations of the input data assuming a normal distribution for statistical evaluation. Models with fixed effects are presented to prevent underestimation of the coefficients. We selected the normal distribution to rule out overestimation of the p-values (i.e. the confidence intervals presented are definitely not too wide). Deviating from Vermeulen et al. [1], we selected mg/m^3-years = 1000 × µg/m^3-years as the exposure unit, to keep the overview clearer. Replicating the analysis as performed in Vermeulen et al. [1], i.e. Steen_Silv_Garsh (adj), results in the highest risk estimate of all data variations. If the modifications of the Silverman case–control study in accordance with Crump et al. 2015 (REC4, REC6, both adjusted for radon exposure) and Garshick et al. [3] are included in the analysis, the risk estimates are far lower (Steen_SilvREC4_Garsh_mod, Steen_SilvREC6_Garsh_mod: the coefficients are reduced roughly by half compared with the Vermeulen analysis). If the Steenland paper, which the authors believe is only exploratory in nature, is excluded, the modified risk estimates do not result in significant risk increases when incorporating Möhner et al. [17] (adapted) in the meta-analysis (SilvREC4_Garsh_mod_Moehn, SilvREC6_Garsh_mod_Moehn). The meta-coefficient is approx. 10 – 20 % of the value found in Vermeulen et al. [1] in the primary analysis.

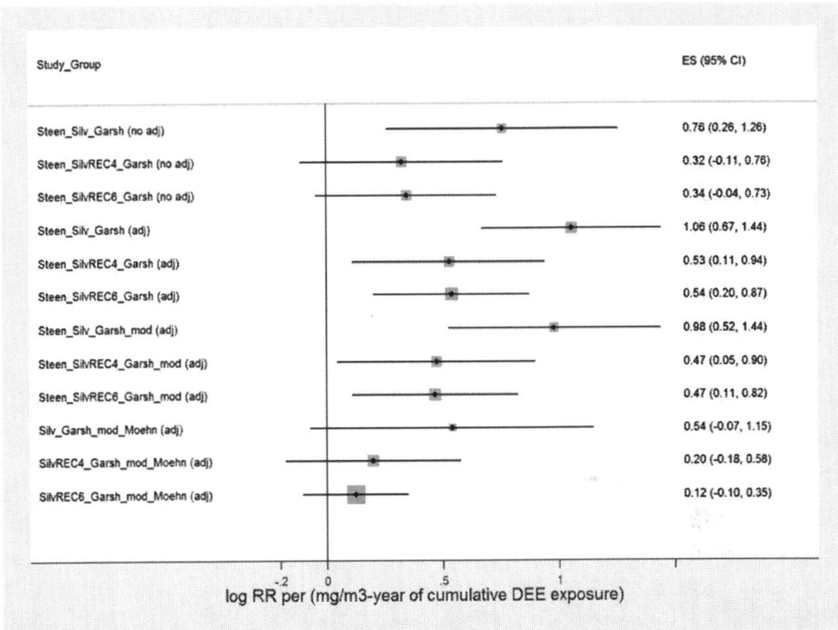

Figure 2. Meta-analyses. Linear regression with fixed effects for log RR with weights proportional to the inverse of the respective variance (the box size illustrates the precision of the meta-estimate). Coefficients (effect size, ES) of the cumulative exposure (Dose) to DEE in mg/m^3-years with a 95 % confidence interval of the coefficient, calculated using the standard normal distribution. Steenland et al. [2] and Silverman et al. [4] were evaluated as used by Vermeulen et al. [1], as well as in accordance with the re-analysis by Crump et al. [14] with adjustment for the radon exposure (REC4, REC6, cf. Table 3), as was Garshick et al. [3] as used by Vermeulen et al. [1], and Garshick et al. [3] (modified) and the data based on Möhner et al. [17] (adapted)

Non-linearity: Threshold search

The primary analysis by Vermeulen et al. [1] was repeated (method: adjusted regression models, fixed effects), but taking a potential threshold into consideration. When evaluated with corrected coefficients in Garshick coefficients (cf. Table 2: Garshick et al. [3], modified), the analysis finds a threshold of 150 μg/m³-years. However, the threshold does not differ statistically significantly from zero.

If we replace Silverman et al. [4] (Table 1) with the results from Crump et al. [14] (Table 4), the analysis also exhibits a threshold. However, this estimate is more difficult to express statistically due to the considerably weaker exposure-response relationship, and exhibits a broad uncertainty (threshold at 90 μg/m³-years, 95 % CI: 0 μg/m³-years to 361 μg/m³-years).

DISCUSSION

Vermeulen et al. [1] published a meta-regression analysis of three major epidemiological US studies [2–4] which analyzed the association between diesel engine exhaust (DEE), based on the cumulative exposure to elemental carbon (EC) in μg/m³-years, and lung cancer mortality. In their meta-analysis, the authors described a statistically significant dose–response relationship and elevated cancer risks, even at very low exposures. The present re-analysis largely succeeded in reproducing the individual and main findings from the published study data. Of all the meta-analyses we performed, the evaluation of the data as used by Vermeulen et al. [1] resulted in the highest risk estimates. However, an investigation of the heterogeneity in the baseline level of risk – Vermeulen et al. [1] do not report on this – resulted in pronounced differences between the three studies (significant at the 5 % level). All three studies should exhibit a uniform baseline level of RR = 1 at a cumulative DEE exposure of 0 μg/m³-years. A joint analysis of the three studies must therefore be viewed critically from a statistical point of view [34]. Other authors in other situations reject combinations of studies even with considerably less pronounced heterogeneity [35, 36]. This uncertainty in the baseline level renders the use of the analysis by Vermeulen et al. [1] a problem for risk estimates in the lower exposure region. Accordingly, the following statement by Vermeulen et al. [1] must be relativised: “Formal tests of heterogeneity of estimates among the studies were of limited value due to the small number of data points for each study."

Correcting the coefficients in Garshick et al. [3] [cf. 13] reduced the heterogeneity in the baseline relative risk levels between the three studies considerably ($P = 0.63$). This correction was indicated as the coefficients used by Vermeulen et al. [1] are incorrectly adjusted [13]: The cumulative exposure was adjusted for the exposure duration it already contains so that the risk coefficient does not estimate the effect of cumulative exposure. Our modification resulted in far lower risk estimates than reported on this study by Vermeulen et al. [1]: The corrected dose coefficient is lower by more than an order of magnitude, and far from being significant ($P = 0.6$). However, the correction of the Garshick coefficients only reduces the meta-coefficient by approx. 7 % and the statistical significance is maintained, although the P-value increases from 0.002 to 0.006.

Another aspect is important for correct evaluation of the Garshick study (see the Letter to the Editor by Morfeld [13], including the answer from the authors). Garshick et al. [3] made a double adjustment for the year of birth, though with different fineness, so that the models do not break down due to collinearity. However, over-adjustments of this type can distort the coefficient estimate. The authors responded that they incorporated the years twice to obtain meaningful results. That was the only way the proportional hazards assumption was fulfilled. However, this explanation does not change the double use of the year of birth information, and the potential over-adjustment it causes.

The highest exposure value included in the meta-study came from the DEMS case–control study by Silverman et al. [4]: 1036 μg/m^3-years with an OR = 2.83 (95 % CI: 1.28 to 6.26). Thus, the US mining study is particularly relevant for the meta-analysis. Critical comments and a list of open questions were published [37] on the DEMS publications [4, 12], to which the authors reacted with a letter to the editor [38]. However, many aspects remain open [see the author answer from Morfeld in 38]. Some of these open questions could be answered by additional analyses, which requires access to the original data. Access to the original data of the DEMS study has only been granted to a few researchers (working groups of S. Moolgavkar [30] and K. Crump [14]). It is currently unclear whether these authors had unrestricted access to all original data. However, aspects of the manner in which the Crump et al. [14] team was given access to the DEMS data is provided in the

"Epilogue" section of their paper. The results of the DEMS re-analysis were presented in a HEI (Health Effects Institute) webinar [28, 29] and published in detail recently [14, 30].

In a Letter to the Editor on Vermeulen et al. [1], Crump [39] reported that an evaluation of the DEMS case–control study with an exposure lag of 5 years leads to far lower meta-risk coefficients than with the original value of 15 years. Crump also pointed out that the other two studies incorporated [2, 3] used a lag of 5 years. The authors qualitatively confirmed the Crump result, but did not consider it relevant, as the adjustment in the DEMS study using a lag of 15 years was better and there are inevitably differences in the exposures recorded in the different studies. Furthermore, Vermeulen et al. [1] referred to a sensitivity analyses they performed, in which they explored the influence a different lagging of the exposure has. However, the note by Crump [39] shows further uncertainties in the Vermeulen meta-analysis, which cannot be determined without access to the original DEMS data. Our re-analysis should be repeated applying different lags to the studies. Unfortunately, the necessary data are not publicly available.

Moolgavkar et al. [30] re-analysed, using both proportional hazard and biological based mechanistic models, the DEMS cohort study by Attfield et al. [12] and pointed out two important aspects:

1) Time-dependent factors are superimposed, so that model coefficients should not be estimated without considering the interaction with age. Accordingly, stating an isolated risk coefficient – as described in Attfield et al. [12] – does not make any sense according to Moolgavkar et al. [30].

2) One mine (limestone mine) is an outlier in the data. The DEE exposures in this mine are the lowest, but the risk estimates are the highest of all the mines. A Cox regression analysis of the data (without excluding the higher exposures as in Attfield et al. [12]) shows that a significant exposure-response association ($P = 0.0014$) was found in the limestone mine alone. If the limestone mine is excluded from the analyses, the P-value in the overall cohort increases from 0.02 to 0.18 [28], i.e., after excluding the limestone mine, no significant dose–response association results between the DEE exposure and lung cancer mortality in the DEMS cohort study. Although this special role of the limestone mine was not confirmed in the case–control re-analysis by Crump et al. [14], this observation about the limestone mine is a major interpretation problem for the cohort study and the publication by

Attfield et al. [12]. It is noteworthy that Moolgavkar et al. [30] described the ventilation in the limestone mine as follows: “All seven of these mines had substantial mechanical ventilation supplying large quantities of air to minimize airborne dust concentrations and, in the case of the trona mines, to minimize the build-up of methane, an explosive gas. The limestone mining operation was quite different from that in the seven other mines. The limestone mine primarily used natural ventilation, with air flowing up or down vertical shafts between the surface and the mining operations.” In addition, the percentage of radon measurements above detection limit was 85 % in the limestone mine, the highest among all 8 mines; the average percentage among the other 7 mines was only 36 % (Roger O. McClellen, personal communication, 27 July 2015). This may explain why Crump and colleagues were unable to reproduce Moolgavkar’s finding of the limestone mine being an outlier: “Moolgavkar et al. could not control for other covariables, including smoking and radon, because these data were not available” (Crump et al. [14]). Such a control for other covariables was done by Crump et al. [14].

Crump et al. [14] re-analysed the DEMS case–control study by Silverman et al. [4] and also investigated the influence of some covariables, which Silverman et al. [4] did not include in their final models. In the original paper, analysing the association of cumulative DEE exposure (lag = 15 years) and lung cancer mortality led to a trend *P*-value of $P = 0.001$ (Table 3, Silverman et al. [4]), which Crump et al. confirmed: $P = 0.0006$ in their re-analysis. Crump et al. [14] (Table 3) reported a trend P-value of $P = 0.02$, if adjustments are also made for radon exposure. This proves that the radon exposure has a considerable influence on the association, and weakens the statistical significance of the DEE variables after taking this covariable into account. Crump et al. [14] also developed six new DEE exposure metrics, as an alternative to the estimates used in Attfield et al. [12] and Silverman et al. [4]. If these alternative DEE exposure metrics are used, and also adjusted for the radon exposure, there is no significant association between the cumulative DEE exposure and lung cancer mortality in any constellation studied ($P \geq 0.17$ or the trend is negative). The analyses by Crump et al. [14] also did not reveal any significant association for the exposure metrics as used in Attfield et al. [12] and Silverman et al. [4], when the individual data in the analysis were used instead of the average values of the groups, and an adjustment was made for radon exposure: $P \geq 0.65$ or the trend is negative. The authors wrote: “Most importantly, we used the radon concentration data for the DEMS cohort provided by the DEMS investigators. When adjustment was made for radon, a known human lung carcinogen, the effect of REC on the association with lung cancer mortality was confined to only the three DEMS REC estimates. Most notably, there was no evidence of an association with any of the six alternate REC estimates, including REC6. When T2 trend tests were conducted, based on the use of individual worker REC estimates, the results were less statistically significant and in many cases the trends were negative. Indeed, for miners who always worked underground, five of the six REC metrics exhibited negative trends."

Moolgavkar et al. [30] and Crump et al. [14] concluded that the DEMS analyses by Attfield et al. [12] and Silverman et al. [4] are not suitable for use on their own in a quantitative risk analysis. The limitations of this DEMS data must also be taken into consideration for meta-analyses and derived limit values. The issue of Risk Analysis that contained the Moolgavkar et al. [30] and Crump et al. [14] papers contained a "note" by the Editors, Cox and Lowrie, on the use of results of analyses of the same data set using alternative models [40]. The Editors described that the reanalysed data set was influential in IARC's decision to classify DEE as a human carcinogen and concluded: "These findings can be viewed as raising important questions about the usefulness and reliability of expert judgements about the causal interpretation of model-dependent associations in general, and about whether DEE is in fact carcinogenic to humans in these studies in particular."

If the paper by Crump et al. [14] (REC4, REC6, both adjusted for radon exposure) is incorporated instead of Silverman et al. [4], our meta-reanalysis of Vermeulen et al. [1] results in a considerably weakened association between DEE exposure and lung cancer mortality. Tables 10, 11, 12 are consistent in showing that a corresponding replacement of the original data roughly halves the dose coefficient in the meta-regression. If we analyse Steenland et al. [2], Crump et al. [14] and Garshick et al. [3] (modified) together, the dose–response association in an adjusted regression model with fixed effects is just about significant ($P = 0.03$), i.e., less certain than in the original data analysis ($P < 0.000001$).

We were unable to apply the recommended Greenland/Longnecker method on the oldest study, that of Steenland et al. [2], as the publication lacks important additional information. However, we did perform meta-analyses of Garshick et al. [3], Silverman et al. [4] and Möhner et al. [17] with this statistical method. Regardless of the evaluation method (random, fixed, Greenland/Longnecker), a joint analysis of Garshick et al. [3] (modified), Silverman et al. [4] and Möhner et al. [17] (adapted) resulted in similar coefficient estimates, whereby the Greenland/Longnecker estimate was between the results from the model with random effects and the adjusted model with fixed effects. This justified focussing our analyses on the adjusted model with fixed effects, which made analyses possible in all data situations. After excluding the explorative study by Steenland et al. [2], the lowest risk estimate resulted in a meta-analysis of the three studies Garshick et al. [3] (modified), Silverman et al. [4] (modified in accordance with Crump et al. [14] and Möhner et al. [17] (adapted). In this evaluation, the meta-coefficient decreased to approx. 10 % to 20 % of the value published by Vermeulen et al. [1] as a result of their primary analysis. In addition to this, the association between the DEE exposure and lung cancer mortality in this analysis is no longer statistically significant.

Vermeulen et al. [1] noted: "We were not able to investigate other model forms in our meta-regression, beyond the linear and spline curves because of the limited number of data points. If nonlinear exposure-response curves were actually a better fit (e.g., attenuation at higher exposures, for which there is some evidence in Silverman et al. (2012), then this might change the estimate burden of disease due to DEE." If we supplement the meta-analysis of the three US studies [2–4]) with a threshold search, i.e., examine the dose–response relationship for non-linearity, and use the corrected Garshick coefficients it results in a threshold estimate for the cumulative DEE exposure at 150 $\mu g/m^3$-years. However, this estimate does not differ significantly from zero.

Sun et al. [41] created an overview of the results from 42 cohort studies and 32 case–control studies on the association between the DEE exposures and lung cancer. The authors concluded: "Overall, neither cohort nor case–control studies indicate a clear exposure-response relationship between DE exposure and lung cancer. Epidemiological studies published to date do not allow a valid quantification of the association between DE and lung cancer." Although this research does not reach the methodological level of the meta-analytical approach by Vermeulen et al. [1], the varying study results in Sun et al. [41] verify the uncertainty underlying epidemiological studies on the association between DEE and lung cancer.

Similarly to our analysis of the Vermeulen analysis, the re-analysis of the German potash miner study by Möhner et al. [17] led to a clearly weakened and different statement compared with the original research [42]: "Only for very high cumulative dose, corresponding to at least 20 years of exposure in the production area, some weak hints for a possible risk increase could be detected."

The meta-regression analysis performed in this paper has considerable restrictions. Major limitations result from the fact that neither DEE concentration values (only details on cumulative exposures) nor the individual data for this research project were available. Threshold analyses for dust should focus on a concentration threshold [see the discussion in 43]. Empirical findings on thresholds in quartz dust exposure, based on the German porcelain worker cohort, did not result in a threshold for the cumulative exposure, but did for the concentration [43]. Similarly, it is probable in this case that the restriction of the data to cumulative DEE exposure means that we can assume that the actual concentration threshold is underestimated. The statistical significance of the finding would be far clearer if the original studies were analysed incorporating concentration values in the meta-analysis. The data are only available in aggregated form (grouped data) (cf. Table 1), while the original data are individual. Analyses of this type are difficult when using data already collapsed into categories, as results can occur depending on how the cut points are chosen, and precision of estimates decreases with categorisation in comparison to continuous analyses. In general, categorisations result in information losses, potential distortions and decreases in power [44].

The limitations mentioned affect the analysis by Vermeulen et al. [1] and our re-analysis accordingly. Although all main authors of the individual studies are co-authors of the meta-analysis, Vermeulen et al. [1] did not pool and analyse the original data, but focused on grouped data from the result tables of the three publications. However, reliable analyses should refer to the individual data and consider the DEE concentration as a key variable in analyses. We note that a similar point was made by Crump [39]: "Vermeulen et al. used very crude exposure summaries (e.g., midpoints of exposure intervals)."

The meta-regressions performed here show significant variations in the results, depending on the study data incorporated or the analytical methods applied. The data used by Vermeulen and colleagues led to the highest risk estimates in our meta-analysis (statistically significant).

After excluding the explorative study by Steenland et al. [2], the lowest risk estimate resulted in an analysis of the three studies Garshick et al. [3] (modified), Silverman et al. [4] (modified in accordance with [14]) and Möhner et al. [17] (adapted). In this evaluation, the meta-coefficient decreased to approx. 10–20 % of the value published by Vermeulen et al. [1] as the main result of their primary analysis. The association between DEE exposure and lung cancer mortality in this analysis is not statistically significant. The risk estimates derived from Vermeulen et al. [1] in the very low exposure range and the corresponding limit value proposals derived [6] are therefore less than convincing, as the data – after correction of the Garshick coefficients – are also generally compatible with a threshold.

Toxicological results of the current ACES study from controlled long-term experiments on rats with new technology diesel exhaust (NTDE) did not exhibit tumour growth or precancerous conditions [45], by contrast to earlier studies with traditional DEE (TDE) from diesel motors without particle reduction and without other forms of exhaust treatment. The summary of IARC workshop results for re-evaluation of DEE in 2012 revealed that only toxicological data with pre-2000 motors and fuel technologies were incorporated [5]. Accordingly, the IARC classification from 2012 referred only to possible carcinogenic potential from diesel engine exhaust without modern exhaust treatment (TDE), using the epidemiological data from Steenland et al. [2], Garshick et al. [3], Silverman et al. [4]. However, the new results from the animal experiments in the ACES study highlight the need to include engine and exhaust treatment technologies in the discussion on workplace limit values and possible lung cancer risks from DEE. McClellan et al. [46] also explicitly mentioned the qualitative and quantitative differences between TDE and NTDE and recommended that these differences should be considered when evaluating the carcinogenic risks. The authors reviewed the substantial changes made in diesel technology, and the resulting changes in diesel exhaust emissions from post-WW II to the present time. The changes in technology and emissions post-1990 have been particularly dramatic. This raises questions with regard to the use of the findings of any of the epidemiological studies analysed in this paper for projecting lung cancer risks of diesel exhaust exposures post-2000.

Regardless of the fundamental restrictions mentioned above, the present re-analysis also revealed that the results of the meta-regression study by Vermeulen et al. [1] should not be used in any quantitative lung cancer risk evaluation without reservations, as the results vary significantly depending on the input data selected and the statistical methods used. This is particularly true for the low exposure region.

CONCLUSIONS

- Vermeulen et al. [1] published a meta-regression analysis of three key epidemiological US studies on the association between diesel engine exhaust (DEE) and lung cancer. They found a statistically significant dose–response relationship and elevated cancer risks even for very low DEE exposures.
- The present re-analysis largely succeeded in reproducing the individual cohort results and main meta-findings of Vermeulen et al. [1] from the published study data. Our meta-regressions, however, show significant variations of the results, depending on the study data incorporated or the analytical methods applied.
- Therefore, the results of the meta-regression analysis by Vermeulen et al. [1] should not be used in a risk assessment without reservation, especially not in the low-DEE exposure range.

DECLARATIONS

Acknowledgements

The research by PD Dr. Peter Morfeld was supported by the European Research Group on Environment and Health in the Transport Sector (EUGT, www.eugt.org). EUGT's goal is to study the effects and interaction of emissions, immissions and health. We wish to thank two unknown reviewers for helpful comments on the manuscript.

AUTHORS' CONTRIBUTION

PM conducted the analysis. MS drafted the introduction and discussion sections. Both authors read, revised and approved the final manuscript.

REFERENCES

1. Vermeulen R, Silverman DT, Garshick E, Vlaanderen J, Portengen L, Steenland K. Exposure-response estimates for diesel engine exhaust and lung cancer mortality based on data from three occupational cohorts. Environ Health Perspect. 2014;122(2):172–7.
2. Steenland K, Deddens J, Stayner L. Diesel exhaust and lung cancer in the trucking industry: exposure-response analyses and risk assessment. Am J Ind Med. 1998;34(3):220–8.
3. Garshick E, Laden F, Hart JE, Davis ME, Eisen EA, Smith TJ. Lung cancer and elemental carbon exposure in trucking industry workers. Environ Health Perspect. 2012;120(9):1301–6.
4. Silverman DT, Samanic CM, Lubin JH, Blair AE, Stewart PA, Vermeulen R, et al. The diesel exhaust in miners study: a nested case–control study of lung cancer and diesel exhaust. J Natl Cancer Inst. 2012;104(11):855–68.
5. Benbrahim-Tallaa L, Baan RA, Grosse Y, Lauby-Secretan B, El Ghissassi F, Bouvard V, et al. Carcinogenicity of diesel-engine and gasoline-engine exhausts and some nitroarenes. Lancet Oncol. 2012;13(7):663–4.
6. Park RM. Diesel engine emissions and risk assessment at NIOSH. In: Presentation at the workshop "Diesel Exhaust, Lung Cancer and Quantitative Risk Assessment". Boston, MA: Health Effects Institute; 2014.
7. The Advisory Committee on Safety and Health at Work: Supplementary opinion on the approach and content of an envisaged proposal by the Commission on the amendment of Directive 2004/37/EC on Carcinogens and Mutagens at the workplace. Supplementary opinion Doc. 727/13. Adopted on 30/05/2013. European Commission, DG Employment, Social Affairs and Inclusion Employment and Social Legislation, Social

Dialogue, Health, Safety and Hygiene at Work. 2013. Available from: https://circabc.europa.eu/d/d/workspace/SpacesStore/85739c10-4b49-42ac-8982-aafc9a38f0eb/Doc%20%20727-13-%20EN%20Supplementary%20CMD%20Opinion_FINAL.pdf. Accessed June 10, 2015

8. Cherrie JW, Gorman M, Shafrir A, van Tongeren M, Mistry R, Sobey M, Corden C, Rushton L, Hutchings S: Health, socio-economic and environmental aspects of possible amendments to the EU directive on the protection of workers from the risk related to exposure to carcinogens and mutagens or work. 2011. Available from: http://ec.europa.eu/social/main.jsp?catId=716&langId=de&moreDocuments=yes. Accessed June 15, 2015

9. Rothman KJ, Greenland S, Lash TL. Modern Epidemiology. 3rd ed. Philadelphia: Lippincott Williams & Wilkins; 2008.

10. Steenland NK, Silverman DT, Hornung RW. Case–control study of lung cancer and truck driving in the Teamsters Union. Am J Public Health. 1990;80(6):670–4.

11. Zaebst DD, Clapp DE, Blade LM, Marlow DA, Steenland K, Hornung RW, et al. Quantitative determination of trucking industry workers' exposures to diesel exhaust particles. Am Ind Hyg Assoc J. 1991;52(12):529–41.

12. Attfield MD, Schleiff PL, Lubin JH, Blair A, Stewart PA, Vermeulen R, et al. The diesel exhaust in miners study: a cohort mortality study with emphasis on lung cancer. J Natl Cancer Inst. 2012;104(11):869–83.

13. Morfeld P. Letter to Garshick et al. 2012: Cox model setup may lead to erroneous conclusions. Environ Health Perspect. 2012;120(10):A382 (Author response A382-A383).

14. Crump KS, Van Landingham C, Moolgavkar SH, McClellan R. Reanalysis of the DEMS nested case–control study of lung cancer and diesel exhaust: Suitability for quantitative risk assessment. Risk Anal. 2015;35(4):676–700.

15. Crump K, Van Landingham C. Evaluation of an exposure assessment used in epidemiological studies of diesel exhaust and lung cancer in underground mines. Crit Rev Toxicol. 2012;42(7):599–612.

16. Stewart PA, Coble JB, Vermeulen R, Schleiff P, Blair A, Lubin J, et al. The diesel exhaust in miners study: I. Overview of the exposure assessment process. Ann Occup Hyg. 2010;54(7):728–46.

17. Möhner M, Kersten N, Gellissen J. Diesel motor exhaust and lung cancer mortality: reanalysis of a cohort study in potash miners. Eur J Epidemiol. 2013;28(2):159–68.

18. Allison PD. Fixed effects regression models. Los Angeles: SAGE; 2009.

19. Cameron AC, Trivedi PK. Microeconometrics. Methods and applications. Cambridge: University Press; 2005.

20. Cameron AC, Trivedi PK. Microeconometrics using stata. revisedth ed. Texas: StataCorp LP, College Station; 2010.

21. Diggle PJ, Heagerty PJ, Liang K-Y, Zeger SL. Analysis of longitudinal data. 2nd ed. New York: Oxford University Press Inc.; 2002.

22. Berlin JA, Longnecker MP, Greenland S. Meta-analysis of epidemiologic dose–response data. Epidemiol. 1993;4(3):218–28.

23. Neter J, Wasserman W, Kutner MH. Applied linear statistical models. Regression, Analysis of variance and experimental designs. 2nd ed. Homewood, Illinois: Richard Dr. Irwin; 1985.

24. Greenland S, Longnecker MP. Methods for trend estimation from summarized dose–response data, with applications to meta-analysis. Am J Epidemiol. 1992;135(11):1301–9.

25. Orsini N, Bellocco R, Greenland S. Generalized least squares for trend estimation of summarized dose–response data. The Stata Journal. 2006;6:40–57.

26. Orsini N, Li R, Wolk A, Khudyakov P, Spiegelman D. Meta-analysis for linear and nonlinear dose–response relations: examples, an evaluation of approximations, and software. Am J Epidemiol. 2012;175(1):66–73.

27. StataCorp. Release 12. Statistical Software. TX: College Station, StataCorp LP; 2013.

28. Crump K, van Landingham C, Moolgavkar S, Mc-Clellan R (2014) Reanalysis of the DEMS nested case-control study of lung cancer and diesel exhaust: suitability for quantitative risk assessment. Health Effects Institute, Boston (Webinar September, 2014)

29. Moolgavkar S (2014) Diesel engine exhaust and lung cancer mortality – temporal factors in exposure and risk. Health Effects Institute, Boston (Webinar September, 2014).

30. Moolgavkar SH, Chang ET, Luebeck G, Lau EC, Watson HM, Crump KS, et al. Diesel engine exhaust and lung cancer mortality: Time-related factors in exposure and risk. Risk Anal. 2015;35(4):663–75.

31. Hudson DJ. Fitting segmented curves whose join points have to be estimated. J Am Stat Assoc. 1966;61(316):1097–129.

32. Jones RH, Molitoris BA. A statistical method for determining the breakpoint of two lines. Anal Biochem. 1984;141(1):287–90.

33. Ulm K. A statistical method for assessing a threshold in epidemiological studies. Stat Med. 1991;10(3):341–9.

34. Morfeld P. Letter to the editor: formaldehyde and leukemia: missing evidence! Cancer Causes Control. 2013;24(1):203–4.

35. Al Khalaf MM, Thalib L, Doi SAR. Combining heterogenous studies using the random-effects model is a mistake and leads to inconclusive meta-analyses. J Clin Epidemiol. 2011;64(2):119–23.

36. Checkoway H, Boffetta P, Mundt DJ, Mundt KA. Critical review and synthesis of the epidemiologic evidence on formaldehyde exposure and risk of leukemia and other lymphohematopoietic malignancies. Cancer Causes Control. 2012;23(11):1747–66.

37. Morfeld P. Diesel exhaust in miners study: how to understand the findings? Journal of Occupational Medicine and Toxicology 2012;7(1):Available from: http://www.occup-med.com/content/7/1/10. Accessed August 10, 2015.

38. 38.Silverman DT, Attfield MD. RE: "Diesel exhaust in miners study: how to understand the findings?" by Peter Morfeld. J Occup Med Toxicol 2013:http://www.occup-med.com/content/7/1/10/comments. Accessed June 15, 2015.

39. Crump K. Meta-analysis of lung cancer risk from exposure to diesel exhaust: study limitations. Environmental Health Perspectives 2014;122(9):A230 (author´s reply: 230-231). http://dx.doi.org/10.1289/ehp.1408482.

40. Cox T, Lowrie K. Editorial: what do we really know about effects of diesel exhaust on lung cancer risk? Risk Anal. 2015;35(4):556.

41. Sun Y, Bochmann F, Nold A, Mattenklott M. Diesel exhaust exposure and the risk of lung cancer-a review of the epidemiological evidence. Int J Environ Res Public Health. 2014;11(2):1312–40.

42. Neumeyer-Gromen A, Razum O, Kersten N, Seidler A, Zeeb H. Diesel motor emissions and lung cancer mortality--results of the second follow-up of a cohort study in potash miners. Int J Cancer. 2009;124(8):1900–6.

43. Morfeld P, Mundt KA, Taeger D, Guldner K, Steinig O, Miller BG. Threshold value estimation for respirable quartz dust exposure and silicosis incidence among workers in the German porcelain industry. J Occup Environ Med. 2013;55(9):1027–34.

44. Royston P, Altman DG, Sauerbrei W. Dichotomizing continuous predictors in multiple regression: a bad idea. Stat Med. 2006;25(1):127–41.

45. HEI Advanced collaborative emissions study (ACES): Lifetime cancer and non-cancer assessment in rats exposed to new-technology diesel exhaust. Health Effects Institute. Boston: MA; 2015. Available from: http://pubs.healtheffects.org/view.php?id=430. Accessed June 15, 2015.

46. McClellan RO, Hesterberg TW, Wall JC. Evaluation of carcinogenic hazard of diesel engine exhaust needs to consider revolutionary changes in diesel technology. Regul Toxicol Pharmacol. 2012;63(2):225–58.

Chapter 6

Bronchial Asthma and COPD Due to Irritants in the Workplace - An Evidence-Based Approach

Xaver Baur, Prudence Bakehe and Henning Vellguth

Institute for Occupational and Maritime Medicine, University Medical Center Hamburg-Eppendorf

ABSTRACT

Background: Respiratory irritants represent a major cause of occupational obstructive airway diseases. We provide an overview of the evidence related to irritative agents causing occupational asthma or occupational COPD.

Methods: We searched MEDLINE via PubMed. Reference lists of relevant reviews were also screened. The SIGN grading system was used to rate the quality of each study. The modified RCGP three-star system was used to grade the body of evidence for each irritant agent regarding its causative role in either occupational asthma or occupational COPD.

Results: A total of 474 relevant papers were identified, covering 188 individual agents, professions or work-sites. The focus of most of the studies and the predominant diagnosis was occupational asthma, whereas occupational COPD arose only incidentally.

The highest level assigned using the SIGN grading was 2+ (well-conducted systematic review, cohort or case–control study with a low risk of confounding or bias). According to the modified RCGP three-star grading, the strongest evidence of association with an individual agent, profession or work-site ("**") was found for 17 agents or work-sites, including benzene-1,2,4-tricarboxylicacid-1,2-anhydride, chlorine, platinum salt, isocyanates, cement dust, grain dust, animal farming, environmental tobacco smoke, welding fumes or construction work. Phthalic anhydride, glutaraldehyde, sulphur dioxide, cotton dust, cleaning agents, potrooms, farming (various), foundries were found to be moderately associated with occupational asthma or occupational COPD ("*[+]").

Conclusion: This study let us assume that irritant-induced occupational asthma and especially occupational COPD are considerably underreported. Defining the evidence of the many additional occupational irritants for causing airway disorders will be the subject of continued studies with implications for diagnostics and preventive measures.

Keywords: Work-related asthma; Occupational asthma Occupational COPD; RADS; Irritant-induced asthma

INTRODUCTION

Bronchial asthma and chronic obstructive pulmonary disease (COPD) are common conditions and are the dominating obstructive airway diseases in the general population.

Work-related asthma (WRA) including irritant-induced occupational asthma (OA)

Occupational asthma is defined as a chronic inflammatory disorder of the airways with recurrent episodes of coughing, wheezing, chest tightness, dyspnea, shortness of breath at rest, and reversible airflow limitations caused by a particular occupational environment [1–3].

The available epidemiological and comparative studies and reviews provide evidence that occupational agents cause 5 – 25% of all asthma cases [1, 4–23]. Besides these evident occupational asthma (OA) cases, there is probably an even larger population of sufferers of work-aggravated asthma [24–26]. The latter population shows an objective worsening of pre-existing asthma or non-occupational asthma that develops in parallel with causative conditions encountered in the workplace (Figure 1).

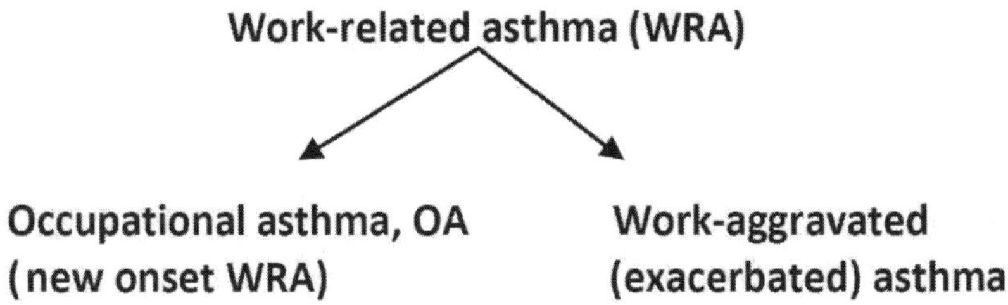

Figure 1. Work-related asthma is divided into occupational asthma and work-aggravated asthma.

Occupational agents eliciting bronchial asthma, i.e. OA, comprise occupational allergens, with their well-defined etiological role and IgE-mediated pathomechanism, as well as occupational agents with unknown pathomechanisms and occupational respiratory irritants, mainly representing low molecular weight chemicals (LMW; <5000 Daltons) causing irritant-induced OA (Figure 2). The latter agents may also elicit occupational COPD (see chapter 1.2) and include chlorine, acids, welding fumes, as well as isocyanates. The etiological role of such low molecular chemicals has not yet been completely clarified, primarily because of the lack of specific diagnostic tests.

There is sparse data available on causes and frequencies of irritant-induced COPD and work-aggravated asthma. Therefore, this work focuses on irritant-induced OA.

There is increasing evidence that irritant-induced OA can be further subdivided into three subcategories as outlined in Table 1 [27–29].

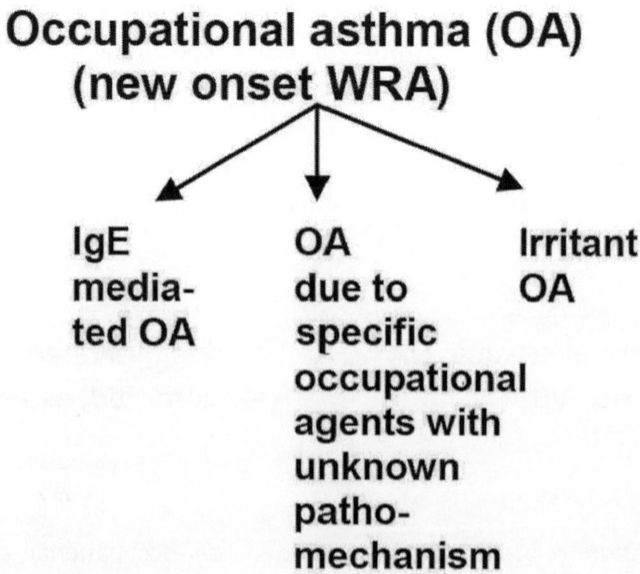

Figure 2. Subgroups of OA.

Table 1. Subcategories of irritant-induced OA

Subcategories of irritant-induced OA	Exposure concentration	Duration of exposure
Reactive airways dysfunction syndrome (RADS)	Extremely high, > OEL	≤1 day
Not so sudden onset of irritant-induced OA	Moderate, around OEL	>1 day <4 months
Low dose irritant-induced OA	Low, below OEL	>4 months

OA = Occupational asthma, OEL = Occupational Exposure Limit.

Many case reports, case series and a few cross-sectional studies demonstrate that a single short-term accidental massive exposure or several short-term high-level exposures to a respiratory irritant can cause asthma within 24 hours without a latency period. Brooks et al. [30] defined this disorder as “reactive airways dysfunction syndrome” (RADS). This term was later extended to irritant-induced OA from multiple, somewhat lower, exposure incidents with a less sudden onset that were also shown to cause this disorder [27, 31–36].

Furthermore, there is evidence that a susceptible subgroup of subjects mainly atopics with non-specific bronchial hyperresponsiveness (NSBHR) suffering from irritant-induced OA, is also affected by chronic exposures to relatively low concentrations of irritant gases, fumes or aerosols [27, 37, 38]. This disorder has been called "low-dose irritant asthma" (or "low-dose RADS"). Corresponding studies indicate respiratory effects including asthma from mainly chronic or repeated exposure to a single irritant or a mixture. Demonstrably causative concentrations of a particular irritant are often below their occupational exposure limits (OELs) or permissible exposure limits (PELs). Such irritant examples include swine confinement facilities [39, 40], exposures to cleaning agents [12, 41], solvents, ozone, endotoxin, formaldehyde, quaternary ammonium compounds, chlorine, bisulfite and SO_2, or acid mist [36, 37, 42–44], diesel exhaust [10, 45, 46], fumigant residues [47], dusts in the textile paper, mineral fiber or construction industries or in mines [48–51], as well as a proportion of cases of potroom asthma [52] and meatwrappers' asthma [53]. Asthma in cold-air athletes may also be relevant [54, 55]. A previous summary of the literature on respiratory effects from asthma due to irritants below their OELs/PELs is available [56]. Many of the earlier exposure limits have been lowered repeatedly in the light of subsequent clinical or epidemiological findings on their respiratory effects. Other limits remain obstinately high given their known irritative effects and/or that they are based on sparse data [56]. Accordingly, adherence to OELs/PELs does not preclude the onset of WRA in susceptible subjects.

The broader definition of these disorders (as used in the legal definition in Germany) includes all irritant-induced obstructive airway diseases irrespective of their causative concentrations and reversibility, i.e. irritant-induced occupational asthma as well as COPD.

Frequency of OA

OA has become the most prevalent occupational lung disease in developed countries [57, 58] and it is one of the most frequent diagnosis among occupational diseases in general [59]. The annual incidence of OA is in the range of 50 per million with extremes up to 250 per million workers and more than 1,300 per million in specific workplaces [57, 60]. As already mentioned there is evidence that occupational agents cause 5 - 25% of of all asthma cases. However, complete registries of OA do not exist and therefore the true frequency of the disease is unknown. Ameille et al. [61] and Fernández-Nieto et al. [59] stated that OA is underestimated among occupational diseases, because many OA cases are not subjected to appropriate diagnostic tests.

Irritant-induced OA is reported to occur in approximately 5 -18% of all OA cases, being the second most common form of OA after allergic OA [36, 62].

Chronic obstructive pulmonary disease (COPD) due to occupational exposure

The diagnosis of COPD is based on chronic productive cough, airflow limitation that is usually not fully reversible, and a progressive, abnormal inflammative response of the lungs mostly caused by long-term smoking and by other noxious particles or gases [1].

During ongoing causative exposures (e.g. smoking particles, droplets and/or gases), airflow limitation is usually progressive and associated with an abnormal inflammatory response of the lungs. Patients with COPD have greater number of neutrophils and alveolar macrophages in bronchoalveolar lavage fluid than healthy non-smokers [63]. Sites of emphysema, which are frequently found in COPD patients, contain large numbers of lymphocytes, and the extent of lymphocyte accumulation correlates with reduction of FEV_1.

In their summaries of the literature, Hnizdo et al. [64], Trupin et al. [65] and Balmes et al. [1] found an occupational contribution in about 15% of COPD cases.

Occupational COPD is identified on epidemiological basis, by observing increased frequencies of COPD among certain working groups [66], e.g. in construction workers [2]. Some occupational exposures (e.g. welding fumes, aluminium, potroom fumes, and cadmium) may cause COPD associated with emphysema [67, 68].

At later stages of OA, the condition of some subjects does not improve over weekends or during holidays and coincides with symptoms of COPD patients. This observation also applies to non-occupational obstructive airways diseases [69, 70] and indicates that a group with changing diagnoses as well as with some overlap between OA and occupational COPD, does exist [66, 71–73].

BACKGROUND AND OBJECTIVE

WRA and occupational COPD are serious and sometimes fatal diseases, which can lead to ill health, inability to work and lost productivity [1, 25, 74–76]. They represent a huge economic burden to the society. For details see Additional file 1: Online Supplement "Economic burden".

The objective of this study is to summarize present knowledge on respiratory irritants causing obstructive airway diseases in humans in the occupational setting and to provide a rating of the strength of evidence for each irritant which has not been previously available.

METHODOLOGY

A systematic review of the literature on occupational irritant-induced OA and occupational COPD due to occupational irritants was conducted. We considered asthma-inducing irritating agents as well as those reported to cause occupational COPD and related disorders, where obstructive ventilation patterns were demonstrated in clinical investigations, cross-sectional studies, cross-shift and/or in long-term exposure studies. Irritating gases mainly occurring in the general environment, such as ozone, and inorganic dusts, including silica, talcum, silicates and other fibers known to cause pneumoconioses, were not considered even though exposure to them is frequently associated with mixed ventilation patterns.

Definitions used

Occupational COPD: chronic bronchitis symptoms and non-reversible airflow limitation due to particular occupational environment (if lung function data was available; otherwise, clinical diagnosis as given by the authors is cited).

Occupational asthma: episodes of shorteness of breath due to particular occupational environment and reversible airflow limitation (if lung function data was available; otherwise, clinical diagnosis as given by the authors is cited).

Obstructive ventilation pattern: we applied reference values of FEV_1/FVC from Brändli, Schindler et al. 2000 [77].

Information sources and selection criteria

Occupational respiratory irritants

To identify the evidence of irritants of the respiratory tract, all agents denoted as "may cause respiratory irritation" by the phrase H335 (previous code R37) and "may cause allergy or asthma symptoms or breathing difficulties if inhaled" H334 (previous code R42) [78] and/or as "irritants" by American Conference of Governmental Industrial Hygienists [79] were initially listed [80]; later this list was compared with results of our database search (see below).

Database search.

We searched for publications reporting investigations exclusively in humans (i.e. animal or in-vitro research was excluded). To be included, the publications had to deal with subjects occupationally exposed to airway irritants.

MEDLINE®-Database was searched with PubMed® from its inception up to December 2007 with the following medical subject headings (MeSH) combinations for each single agent:

"*Agent*"[MeSH] AND "Humans"[MeSH] AND (("Asthma"[MeSH] OR "Asthma/chemically induced"[MeSH] OR "Asthma/immunology"[MeSH]) OR "Pulmonary Disease, Chronic Obstructive"[MeSH] OR "Lung Diseases, Obstructive/*chemically induced"[MeSH] OR "Respiratory Function Tests"[MeSH]) AND ("Accidents, Occupational"[MeSH] OR "Occupational Exposure"[MeSH] OR "Occupational Diseases"[MeSH] OR "Occupational Diseases/chemically induced"[MeSH])).

If more than 20 publications per agent were found, the search was more specified:

"*Agent*"[MeSH] AND "Humans"[MeSH] AND ("Cohort Studies"[MeSH] OR "Case–control Studies"[MeSH] OR "Case–control Studies"[All Fields] OR "Longitudinal Studies"[MeSH] OR "Longitudinal Studies"[All Fields] OR "Cross-Sectional Studies"[MeSH] OR "Cross-Sectional Studies"[All

Fields] OR "Epidemiologic Studies"[MeSH] OR "Epidemiologic Studies"[All Fields] OR "Case Reports"[Publication Type] OR "Meta-Analysis"[MeSH] OR "Meta-Analysis"[All Fields]) AND "adverse effects"[Subheading] AND (("Asthma"[MeSH] OR "Asthma/chemically induced"[MeSH] OR "Asthma/immunology"[MeSH]) OR "Pulmonary Disease, Chronic Obstructive"[MeSH] OR "Lung Diseases, Obstructive/*chemically induced"[MeSH] OR "Respiratory Function Tests"[MeSH]) AND ("Accidents, Occupational"[MeSH] OR "Occupational Exposure"[MeSH] OR "Occupational Diseases"[MeSH] OR "Occupational Diseases/chemically induced"[MeSH])).

Reference list screening

We also considered references in the identified already existing 13 systematic reviews or overviews of causes of work-related asthma or COPD and tried to combine results of both approaches.

Occupational diseases statistics

Further, we considered the following occupational diseases statistics based either on statutory surveillance or registration systems: SWORD 1994–1997 [81–83]; SHIELD 1993 [74]; SORDSA 2001 [84]; SENSOR 2003 [85]; Dokumentation der Berufskrankheiten 2007 (BK-DOK) [86].

Conditions

Four different conditions were accepted for inclusion:

1. Irritant-induced OA including RADS. Asthma caused by single or multiple occupational exposures to airway irritants; de novo irritant-induced OA. Asthma within 24 hours without a latency period caused by short-term high-level exposures to a respiratory irritant known as acute irritant-induced asthma, or as RADS [30].

2. Work-aggravated (exacerbated) irritant-induced OA. Pre-existing or concurrent asthma worsened by work factors [24]. Subjects with work-related asthmatic symptoms, if not differentiated whether new-onset or work-aggravated.

3. Occupational COPD. On epidemiological basis, identified by observing increased frequencies of COPD among certain working groups [66].
4. Obstructive ventilation pattern. Studies about irritant agents, where obstructive ventilation patterns in occupational settings were reported.

Methodological selection criteria

Publications with one of the following study designs were included: Systematic reviews of cohorts, case–control or cross-sectional studies, cohort studies (prospective/retrospective), case–control studies, cross-sectional studies, surveys.

Non-analytic1[a] studies (i.e. case series, follow-up of cases or case reports) were only included when for an agent no studies with one of the above mentioned designs had been identified.

Publications were included when they met any of the following criteria: examining the frequency of irritant-induced OA or asthmatic work-related symptoms in occupationally exposed groups or individuals, reporting the causative role of the specific agent or mixture of agents for irritant-inducing WRA or COPD.

Studies were included when they applied any of the following diagnostic tools: description of work-related asthmatic symptoms (questionnaire), lung function test (LFT), testing for non-specific bronchial hyperresponsiveness (NSBHR) by means of methacholine, histamine or other pharmacological agents, serial spirometry or expiratory peak flow (PEF) monitoring or supervised exposure testing in the workplace, challenge with the help of lung function measurements (SFT), specific inhalation challenge testing (SIC), clinical diagnosis of OA by an expert (occupational or pulmonary physician), and exposure to an irritant agent.

Publication period: No restriction for publication dates were made, last updates were between 5thand 15th June 2012.

Language: English, German, Spanish, Italian or French.

Methodological studies, e.g. on effects of study design and subsequent procedures, and studies with non-occupational disorders were excluded. Publications about occupational agents which do not have an irritant effect on the respiratory tract (e.g. about IgE-sensitizing agents) or with unrelated issues (e.g. studies on immunological questions) were also excluded.

Assessment of study quality

The principal study characteristics and study results were systematically extracted using an extraction sheet (see Additional file 2: Table S1A of online supplement "Methodology").

We assessed study quality with the help of a check list (see Additional file 2: Table S1B of online supplement "Methodology"). The evidence level of each study was graded according to the revised Scottish Intercollegiate Guidelines Network (SIGN) grading system [87]. Since population-based randomized assignment to different levels of irritant exposure are unethical, no randomised controlled trials (RCTs) could be expected on this topic and, thus, no level 1 evidence (as defined by the revised SIGN grading system) [87] would be available. In order to achieve more differentiation among lower evidence grades, we modified the SIGN grading system and added an additional grade (3+) (see Additional file 2: Table S1C of online supplement "Methodology").

Details of the modified RCGP [88] grading system are given in the online supplement "Methodology", Additional file 2: Table S1D.

RESULTS

Overview on publications retrieved

The database search (MEDLINE/ PubMed) yielded 383 potentially relevant publications. 480 additional potentially relevant publications were retrieved from the reference lists of 13 systematic reviews or overviews [2, 30, 36, 80, 89–97], from occupational diseases routine statistics (SWORD 1994–1997 [47, 81–83]; SHIELD 1993 [74]; SORDSA 2001 [84]; SENSOR 2003 [85]; BK-DOK 2007 [86], and from the library of the Institute of Occupational Medicine, Hamburg. Alltogether, the different search approaches yielded a total of 474 relevant studies, including an extreme early study from the year 1932 [98]. (See selection flow diagram, Figure 3).

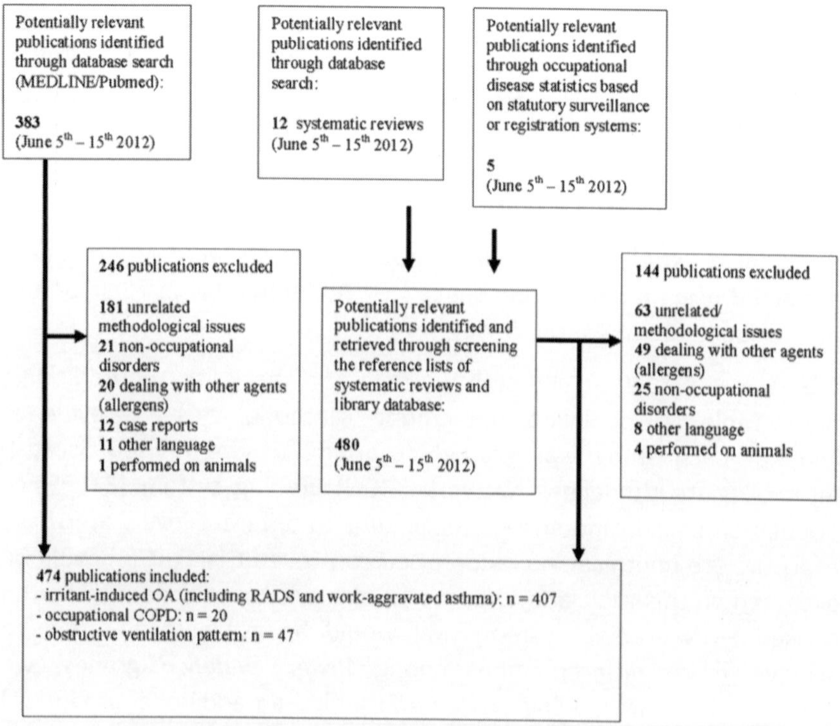

Figure 3. Flow diagram demonstrating the source of references.

Most (n = 337) of the 474 relevant publications were identified through hand searching (i.e. reference list checking of systematic reviews and from our library database).

The 474 publications refer to 131 individual agents, 46 to "mixed" agents and 11 to work-sites or professions reported to cause OA and/ or occupational COPD.

Diagnostic aspects

Many different ways of confirming irritant-induced OA were used, with specific inhalation challenge (SIC) and lung function tests (LFT) being the most reliable diagnostic aids.

SIC was used as the "gold standard" in confirming OA mainly in non-analytical studies (n = 189 studies, i.e. 72% of non-analytical studies). Only few (n = 16, i.e. 7.5% of analytical studies) cohort or case–control reported diagnostic confirmation with SIC.

Another frequently used (n = 191) diagnostic method for OA or occupational COPD was lung function testing (LFT); showing an obstructive ventilation pattern and/or NSBHR related to occupational exposures, mostly in combination with WRA symptoms.

Exclusively self-reported asthma symptoms or physician reported asthma as documented in questionnaires as an alternative diagnosis for OA was used in 36 studies.

Other studies (n = 44) had not clear diagnosis of OA or occupational COPD but reported obstructive ventilation pattern. The number of subjects with asthma symptoms and frequencies of obstructive ventilation patterns and/or NSBHR are provided for each study (see Additional file 3: Table S2E of online supplement "Results").

Irritant-induced OA as outcome

Irritant-induced OA was the focus of most studies and was the predominant diagnosis.

RADS, as a subgroup of irritant-induced OA, was reported to be due to 47 different agents, with the most prevalent being the World Trade Center disaster in 2001 (n = 7 studies), chlorine (n = 11), cleaning agents (n = 18) and isocyanates (n = 46). These were followed by disorders, caused by metam sodium (n = 17), ammonia (n = 11), diesel exhaust (n = 10), acids (n = 9), solvents (n = 8), sulfur dioxide (n = 7), dinitrogen tetraoxide (n = 6), hydrogen chloride (n = 4), smoke (fires, pyrolysis products) (n = 4), chlorofluorocarbons (n = 4), spray paint (n = 3), tear gas (n = 3), bromine (n = 2), dichlorvos (n = 2), sodium azide (n = 2), acrylates (n = 1), amprolium hydrochloride (n = 1), phthalic anhydride (n = 1), bromochlorodifluoromethane (n = 1), bromotrifluoromethane (n = 1), chloramine T (n = 1), chromate (n = 1), hydrazine (n = 1), hydrogen fluoride (n = 1), methylmercaptan (n = 1), phosgene (n = 1), uranium hexafluoride (n = 1), airbag content (n = 1), bleaching agent (n = 1), floor sealant (n = 1), fumigant (n = 1), metal coat remover (n = 1), metal oxide fume (n = 1), pesticides (n = 1), refractory ceramic fibers (n = 1), swine confinement (n = 1).

The majority of asthma-inducing agents elicited OA after prolonged exposure and rarely after a single exposure.

Work-aggravated asthma was of less importance in the literature and occurred in only a few studies [30, 99–104].

Occupational COPD as outcome

Ten agents and five professions or work-sites were reported to cause occupational COPD, as shown in Table 2.

Table 2. Agents and professions showing evidence of occupational COPD

Agents, number of studies (n)	Work-sites or professions, number of studies (n)
o ammonia (1)	o construction work (3)
o cement dust (4)	o swine confinement (1)
o chlorine (1)	o farming (1)
o cleaning agent (1)	o foundry (1)
o mustard gas (1)	o metallurgical industry workers (1)
o diesel exhaust (2)	
o environmental tobacco smoke(1)	
o isocyanate (1)	
o smoke (1)	
o sulphur dioxide (1)	

COPD = Chronic obstructive pulmonary disease.

As already mentioned, occupational COPD was not specifically addressed in most of the studies. Some describe respiratory symptoms, such as chronic bronchitis (n = 21), which may be indicative of COPD. One of the few studies which specifically focused on COPD, was a large retrospective cohort study on diesel exhaust which caused a significantly increased COPD mortality in railroad workers after the introduction of diesel engines in 1945 [105, 106]. Construction work was identified as a cause of occupational COPD in 2 publications [8, 107].

Evidence level of the literature

Some publications investigated more than one irritant agent and thus have been considered several times in our study.

*262*of the 474 publications were non-analytic studies and were rated according to SIGN as 4, 3 or 3+ and consisted of case reports (n = 228), case series (n = 63), and occupational diseases statistics (n = 33) and reviews of that kind of studies (n = 7). The other publications reported analytical studies and were rated according to SIGN as 2+ (n = 15), 2- (n = 103), or 3+ (n = 83).

The highest level was 2+, indicating a well conducted analytical study (case control or cohort studies) with a low risk of confounding or bias (n = 15 studies). Other studies with a similar design had a higher risk of confounding or bias and were individually rated lower by SIGN grading of 2- (n = 30 studies). Most of the other analytical studies were rated with a SIGN grade of 2-, because their design (cross-sectional or longitudinal study) was limiting (n = 82 studies). Cross-sectional studies or longitudinal studies, e.g. those with high risk of confounding or bias, were rated even lower with 3/ 3+ (n = 35 studies). A couple of study designs were difficult to classify epidemiologically, including those which were surveys, mostly with very low analytical evidence, rated 3/ 3+ (n = 53 studies), or larger surveys with a lower risk of confounding or bias, which were graded with 2- (n = 4 studies).

Investigations involving dose–response relationship as a form of scientific evidence were performed in 30 out of 474 studies analyzed [68, 105, 106, 108–133].

Another assessment of the level of evidence found in individual studies is to consider their OR for irritant-induced OA or occupational COPD; this was done in 39 publications [15, 23, 44, 48, 105–108,113, 117–120, 122, 126, 128, 134–156].

Strength of evidence per agent, work-site or profession

The outcome for each agent causing OA or occupational COPD was graded according to the modified RCGP three-star system to classify the strength of evidence of its causative role in irritant-induced OA/ occupational COPD. The strongest evidence achieved was two stars "**" (indicating a moderate strength of evidence provided by generally consistent findings in fewer, smaller or lower quality scientific studies) for 17 (mixed) agents, work-sites or professions. For six of them (chlorine, platinum salts, environmental tobacco smoke, welding fumes, construction work, World Trade Center disaster in 2001), this level was based on well- conducted studies with low risk of confounding and/or bias (SIGN 2+). For eleven of these 17 (mixed) agents, SIGN levels of individual studies were lower (benzene-1, 2, 4-tricarboxylic acid-1,2-anhydride [trimellitic anhydride], cobalt, isocyanates, cement dust, grain dust, animal farming (pig, beef/veal, dairy, poultry), or swine confinement.

Table 3. Overview of individual agents causing irritant-induced OA or occupational COPD

Agent CAS	Strength of evidence (modified RCGP three star grading)	Number of studies per agent	References
Acids			
", acetic	*	3	[108,157,158]
64-19-7			
", not specified	(*)	2	[30,37]
", dodecanedioic	-	1	[159]
693-23-2-1			
", various	-	1	[35]
", " (hydrochloric, hydrofluoric, nitric, perchloric, sulfuric)	-	1	[160]
", hydrochloric	(*)	6	[35,99,161,162]
7647-01-0			
", hydrofluoric	-	1	[163]
7664-39-3			
", sulfuric	*	3	[99,109,164]
7664-93-9			
Acrylates			
", not specified	-	1	[165]
", alkyl cyanoacrylates	(*)	4	[166-169]
", cyanoacrylate glue	-	1	[170]
", " (loctite)	*	4	[44,169,171,172]
53858-53-0			
", methacrylates	*	1	[134]
", methyl 2-cyanoacrylate	-	3	[166,169,173]
137-05-3			
", methylmethacrylate	-	2	[169,174]
80-62-6			
Aluminum salts	*	1	[100]
aluminum fluoride: 7724-18-1			
aluminum sulfate: 10043-01-3			
2-Aminoethanol [2-ethanolamine]	-	1	[175]
141-43-5			
Amino-ethyl-ethanolamine	-	1	[176]
111-41-1			
3-Amino-5-mercapto-1,2,4-triazole	(*)	1	[177]
16691-43-3			
Ammonia	*	6	[41,178-182]
7664-41-7			
Ammonium chloride (triple salt)	-	1	[183]
12125-02-9			
Ammonium thioglycolate	-	1	[184]
5421-46-5			
Amprolium hydrochloride	-	1	[185]
137-88-2			

Anhydrides			
*, various	*	2	[186,187]
*, dioctyl phthalate 117-81-7	-	1	[188]
*, hexahydrophthalic 37226-48-5	-	1	[189]
*, himic 2746-19-2	-	1	[190]
*, maleic 108-31-6	-	2	[191,192]
*, methyltetrahydrophthalic 26590-20-9	-	1	[193]
*, phthalic anhydride 85-44-9	*[*]	5	[194-198]
*, pyromellitic dianhydride 89-32-7	-	1	[199]
*, tetrachlorophthalic anhydride 117-08-8	*	4	[200], [201], [202], [203]
*, benzene-1, 2, 4- tricarboxylic acid 1,2-anhydride [trimellitic anhydride] 552-30-7	**	5	[197,204-207]
Aziridine, polyfunctional 64265-57-2	(*)	2	[208,209]
Azobisformamide [azodicarbonamide] 123-77-3	*	5	[210-214]
Benzalkonium chloride (fumes) 8001-54-5	-	2	[215,216]
1, 2-Benzisothiazoline-3-one (fumes) 2634-33-5	-	1	[217]
Bisulfite, *SO2* *SO2: 7446-09-5*	-	1	[37]
Bromine, hydrobromic acid	-	1	[218]
Bromochlorodifluoromethane (Halon 1211) 353-59-3	-	1	[101]
Bromotrifluoromethane (Halon 1301) 75-63-8	-	1	[219]
Cadmium (fumes) 7440-43-9	*	4	[68,220-222]
Calcium carbonate [chalk powder]	-	1	[110]
Calcium oxide 1305-78-8	-	1	[35]
Captafol (chlorinated thiocarboximide fungicide) 2425-06-1	-	1	[223]
Carbon black dust 1333-86-4	*	1	[224]
Chloramine T (powder dust) 7080-50-4	(*)	5	[225-229]

Chlorhexidine 55-56-1	-	**1**	[230]
Chlorine 7782-50-5	**	**11**	[35,165,231-239]
Chromate (not specified) [see also cement]	(*)	**9**	[240,241] [98,242-247]
Cobalt 7440-48-4	**	**15**	[74,113,247-259]
3-(Diamino-amino)propylamine 3-(dimethylamino) propylamine] 109-55-7	(*)	**1**	[260]
Diamine, aliphatic + cycloaliphatic (hardener) 2855-13-2 (isophorone diamine)	-	**1**	[261]
Diazonium tetrafluoroborate 14239-22-6	-	**2**	[262,263]
Dichlorodiethyl sulfide [mustard gas] +505-60-2	(*)	**1**	[264]
Dichlorvos (organophosphate) 62-73-7	-	**2**	[265,266]
Diethanolamine 111-42-2	-	**1**	[267]
2-Diethylaminoethanol [diethyl aminoethanol] 100-37-8	(*)	**1**	[268]
2-Dimethylaminoethanol [dimethyl ethanolamine] 108-01-0	-	**2**	[269,270]
Diinitrogen tetraoxide [dinitrogentetroxide] 10544-72-6	(*)	**1**	[271]
Ethylenediamine [ethylene diamine] 107-15-3	*	**6**	[168,184,272-275]
Ethylene oxide 75-21-8	-	**2**	[276,277]
Formaldehyde (gas, dust) 50-00-0	*	**9**	[278-283], [284,285]
Freon, (freon-22)	-	**2**	[286,287]
Glutaraldehyde [glutaral] 11-30-8	*[*]	**9**	[74,83,288-294]
Hexachlorophene 70-30-4	-	**1**	[295]
Hexamethylenetetramine 100-97-0	*	**3**	[184,296,297]
Hydrazine 302-01-2	-	**1**	[30]
Iridium salt	-	**1**	[298]
Isocyanates, isocyanurate			
", various (HDI, MDI, TDI)	**	**11**	[57,83,148,149, 165,281,299-303]
", Diphenylmethane diisocyanate [MDI] 5873-54-1	**	**7**	[304-310]
", " prepolymers	-	**1**	[311]
", Hexamethylene diisocyanate [HDI]; plus isodurane Diisocyanate	(*)	**3**	[312,313]

822-06-0			
", HDI biuret plus			[314]
4035-89-6			
", 3-Isocyanatomethyl-3,5,5-trimethylcyclohexyl isocyanate [isophorone diisocyanate, IPDI]	-	1	[315]
4098-71-9			
", Methyl isocyanate [MIC]	*	6	[316-321]
624-83-9			
", 1,5-Naphthylene diisocyanate [NDI]	(*)	[46]	[322-324]
3173-72-6			
", Polymethylene polyphenyl isocyanate	*	1	[325]
9016-87-9			
", Toluene diisocyanate, TDI 2,4: 584-84-9; 2,6:91-08-7	**	12	[35,125,165, 326-334]
", Triglycidil isocyanurate	-	1	[335]
2451-62-1			
", Triphenylmethane triisocyanate	-	1	[336]
Isothiazolinone	-	1	[337]
55965-84-9			
Lauryl dimethyl benzyl ammonium chloride	-	1	[338]
139-07-1			
Metam sodium [methyldithiocarbamate]	-	1	[102]
144-54-7			
Methylmercaptan	-	1	[165]
74-93-1			
Monoethanolamine	-	1	[184]
141-43-5			
N-methylmorpholine	[*]	1	[339]
109-02-4			
Nickel sulphate	(*)	5	[246,340-343]
→anhydrous 7786-81-4			
→hexahydrate 10101-97-0			
Ninhydrin 485-47-2	-	1	[335]
Nitrogen chloride [nitrogen trichloride, trichloramine]	[*]	2	[150,344]
10025-85-1			
Ozone (gassings)	*	1	[345]
10028-15-6			
Palladium	-	1	[346]
7440-05-3			
Paraphenylenediamine	(*)	1	[347]
106-50-3			
Paraquat	*	2	[128,151]
4685-14-7			
Persulfate			
", not specified	(*)	2	[348,349]
", ammonium	-	1	[350]
", potassium (7727-21-1) and ammonium peroxydisufate (7727-54-0)	[*]	5	[351]

*, alcalic	-	1	[352]
*, Sodium persulfate	-	1	[353]
7775-27-1			
*, Dipotassium peroxo-peroxodisulfate [potassium persulfate] 7727-21-1	-	1	[354]
*, Diammonium peroxodisulfate [ammonium persulfate]	*	4	[355-357]
7727-54-0			
Phenylglycine acid chloride	*	1	[358]
39478-47-2			
Phosgene	-	2	[35,359]
75-44-5			
Piperazine dihydrochloride	*	3	[130,176,274]
142-64-3			
Platinum salts	**	8	[131,176,360-365]
(7440-06-4)			
Polyethylene	-	3	[366-368]
9002-88-4			
Polymethyl-methacrylate [plexiglas powder]	-	1	[369]
9011-14-7			
Polypropylene, heated to 250 °C	[*]	2	[370,371]
9003-07-0			
Polyvinyl chloride (fume)	*	8	[53,372-376] [377,378]
9002-86-2			
Potassium dichromate	(*)	1	[379]
7778-50-9 (see also chromium;cement)			
Potassium aluminum tetrafluoride	(*)	1	[380]
14484-69-6			
Rosin core solder, thermal decomposition [colophony]	*	6	[74,83,381-383]
8050-09-7			
Sodium azide (powder dust)	-	1	[384]
26628-22-8			
Sodium iso-nonanoyl oxybenzene sulphonate [SINOS] 123354-92-7	(*)	3	[385-387]
Sodium metabisulfite [metabisulfite sodium]	(*)	6	[103,388-392]
7681-57-4			
Styrene monomer	(*)	3	[132,393,394]
100-42-5			
Sulfur dioxide	*[*]	5	[35,154,395-397]
7446-09-5			
Sulfathiazole	-	1	[398]
72-14-0			
Terpene (3-carene)	-	2	[399,400]
13466-78-9			
Tetrachloroisophthalonitrile (fungicide)	-	1	[401]
Tetrahydrothiophene	-	1	[402]
110-01-0			
Tetramethrin [1-(5-tretrazoly)- 4-guanyl-tetrazene hydrate] 7696-12-0	-	2	[338,403]
Tributyl tin oxide [carpet fungicide]	-	1	[404]

Triethanolamine 102-71-6	-	1	[175]
Tungsten carbide 11130-73-7	-	1	[405]
Tylosin tartrate	-	1	[406]
Uranium hexafluoride 7783-81-5	-	2	[30,407]
Urea (fume) 57-13-6	-	1	[104]
Urea formaldehyde foam 64869-57-4/ **Phenol-formaldehyde resin** 9003-35-4	-	1	[408]
Vanadium 7440-62-2 + **divanadium pentoxide** 1314-62-1	*	5	[84,409-412]
Zinc (fume) 7440-66-6	-	3	[413-415]
Zinc chloride (fume) 7646-85-7	-	1	[183]
Mixed agent	**Evidence level**	**Number of studies per agent**	
Acid fluxes	-	1	[74]
Acrylic acid	-	1	[166]
Airbag content	-	1	[416]
Bleaching agent (fumes)	-	1	[99]
Cement 65997-15-1	**	14	[111,133,235, 417-427]
Chlorofluorocarbons (degradation products)	(*)	2	[428,429]
Cleaning agents (not specified)	*[*]	9	[15,37,41,112, 135,430-433]
", detergents	-	2	[434,435]
Coffee, green	*	5	[436-440]
Cotton (dust, raw) CNT 750	*[*]	(12)	[48,114-116,42: 441-447]
Cutting oil	-	2	[37,448]
Diesel exhaust	*	5	[10,45,46,105,10
ECG ink	-	2	[449,450]
Endotoxin (see also cotton dust, swine confinement, poultry confinement, house dust)	*	2	[41,451]
Environmental tobacco smoke	**	10	[117,118,138-14 452,453]
Floor sealant (aromatic hydrocarbons)	-	1	[30]
Fumigating agent	-	1	[30]
Furan-based binder	-	1	[454]
Grain	**	9	[48,122-124, 455-459]
", rice	[*]	1	[460]
Hairdressing chemicals	(*)	1	[281]

Lubricants (not specified)	(*)	2	[10,57]
Metal coat remover (coating removing chemical)	-	1	[30]
Metal oxide (fume)	-	1	[461]
Metal working fluids [MWF]	-	1	[462]
Oil (spill)	*	1	[126]
Paint (fumes)	*	4	[35,127,463,464]
Paper dust A111	(*)	1	[48]
Perfume agents (research lab)	-	1	[37,465]
Pesticides (not specified)	*[*]	5	[129,152,266, 466,467]
Polyamines, aliphatic	[*]	1	[468]
Polyester	(*)	2	[469,470]
Potroom aluminum smelting	*[*]	10	[52,247,471-478]
Powder paints	(*)	1	[479]
Pyrazolone (see reactive dye)	-	1	[480]
Reactive dyes	*	5	[481-485]
Refractory ceramic fibers [RCF]	*	2	[486,487]
Smoke (fires, pyrolysis products)	*[*]	5	[30,41,488-490]
", (oil fire and dust storm)	*	1	[491]
", (biomass, indoor)	(*)	1	[492]
Soldering flux (fumes)	*	4	[183,493-495]
Solvents (not specified)	*	4	[10,48,496,497]
Spray paint	-	1	[30]
Tall oil	-	1	[498]
Tear gas	-	4	[499-502]
Welding fumes	**	18	[41,48,57,83,247, 281,464,503-513]
Work-site or profession	**Evidence level**	**Number of studies per work-site or profession**	
Ceramic production	*[*]	2	[514,515]
Cleaners			
Construction work (dust, agent not specified)	**	5	[8,107,136, 516,517]
Farming			
", (various)	*[*]	9	[119,120,144-146,518-521]
", animals (pig, beef/veal, dairy, poultry)	**	3	[522-524]
Foundry [see also isocyanates (MDI)]	*[*]	4	[121,147,525,526]
Health care workers	*[*]	4	[23,527-529]
Metallurgical industry workers	*	1	[530]
Poultry confinement	*(*)	4	[153,531-533]
Poultry confinement, slaughtery house	*	3	[534-536]
Swine confinement	**	8	[39,40,141,155, 537-540]
World Trade Center disaster 2001	**	8	[133,156,541-546]

[ ] down-grading due to lower quality of clinical investigations relative to the scale of the scientific level of the study.

(*) up-grading due to at least 5 cases without contradictory findings.

Table 4. Strength of evidence for agents, professions and work-site according to the modified RCGP three-star system **[88]**

Evidence level (modified RCGP three-star grading)	Number of agents/work-sites or professions	Agents, work-site or profession [Synonym] (CAS)
***	0	-
**	17	Benzene-1, 2, 4-tricarboxylic acid-1,2-anhydride [trimellitic anhydride] (552-30-7); chlorine (7782-50-5); cobalt (7440-48-4); various isocyanates, isocyanurate (HDI, MDI, TDI), diphenylmethane diisocyanate [MDI] (5873-54-1), toluene diisocyanate, TDI 2,4 (584-84-9), TDI 2,6: (91-08-7); platinum salts (7440-06-4); cement ; environmental tobacco smoke; grain ; welding fumes; construction work (dust, agent not specified); farming, animals (pig, beef/veal, dairy, poultry); swine confinement; World Trade Center disaster 2001
*[*]	12	Ceramic production; Phthalic anhydride (85-44-9); glutaraldehyde [glutaral] (11-30-8); sulfur dioxide (7446-09-5); cotton (dust, raw) CNT 750; potroom aluminum smelting; farming (various); foundry; smoke (fires, pyrolysis products); pesticides (not specified); cleaning agents (not specified); health care workers
*	39	Acetic acid (64-19-7); sulfuric acid (7664-93-9); metacrylates, loctide® (53858-53-0); aluminum salts [aluminum fluoride] (7724-18-1); aluminum sulfate: (10043-01-3); ammonia (7664-41-7); various anhydrides; tetrachlorophthalic anhydride (117-08-8); azobisformamide (123-77-3); cadmium (fumes) (7440-43-9); carbon black dust (1333-86-4); ethylenediamine (107-15-3); formaldehyde (gas, dust) (50-00-0); hexamethylenetetramine (100-97-0); methyl isocyanate [MIC] (624-83-9); naphthylene diisocyanate (3173-72-6); polymethylene polyphenyl isocyanate (9016-87-9); N-methylmorpholine (09-02-4); ozone (gassings) (10028-15-6); paraquat (4685-14-7); diammonium peroxodisulfate (7727-54-0); phenylglycine acid chloride (39478-47-2); piperazine dihydrochloride (142-64-3); polyvinyl chloride (fume) (9002-86-2); rosin core solder; thermal decomposition (8050-09-7); vanadium (7440-62-2) + divanadium pentoxide (1314-62-1); cleaning agents (not specified); green coffee ; diesel exhaust; endotoxin; oil (spill); paint (fumes); pesticides (not specified); reactive dyes; refractory ceramic fibers [RCF]; smoke (fires, pyrolysis products; oil fire and dust storm); soldering flux; solvents (not specified); health care workers; poultry confinement; slaughtery house; metallurgical industry workers
[*]	3	Nitrogen chloride (10025-85-1); polyamines, aliphatic; potassium persulfate (7727-21-1) and ammonium peroxydisufate (7727-54-0); grain rice
(*)	29	Acids not specified; hydrochloric acids (7647-01-0); alkyl cyanoacrylates; 3-amino-5-mercapto-1,2,4-triazole l(16691-43-3); aziridine, polyfunctional (64265-57-2); chloramine T (powder dust) (7080-50-4); chromate (not specified); 3-(diamino-amino)propylamine (109-55-7); dichlorodiethyl sulfide (505-60-2); 2-diethylaminoethanol (100-37-8); dinitrogen tetraoxide (10544-72-6); hexamethylene diisocyanate [HDI], plus isodurane diisocyanate (822-06-0); HDI biuret plus (4035-89-6); nickel sulphate anhydrous (7786-81-4); hexahydrate (10101-97-0); paraphenylenediamine (106-50-3); persulfate (not specified); polypropylene, heated to 250 ℃ (9003-07-0); potassium dichromate (7778-50-9); potassium aluminum tetrafluoride (14484-69-6); sodium iso-nonanoyl oxybenzene sulphonate [SINOS] (123354-92-7); sodium metabisulfite (7681-57-4); styrene monomer (100-42-5); chlorofluorocarbons (degradation products); hairdressing chemicals; lubricants (not specified); paper dust A111; aliphatic polyamines; polyester; powder paints; smoke (biomass, indoor)
-	93	Acids various; dodecanedioic (693-23-2-1); hydrofluoric acids (7664-39-3); cyanoacrylate glue; methyl 2-cyanoacrylate (137-05-3); methylmethacrylate (80-62-6); 2-aminoethanol (141-43-5); amino-ethyl-ethanolamine (111-41-1); ammonium chloride (triple salt) (12125-02-9); ammonium thioglycolate (5421-46-5); amprolium hydrochloride (137-88-2); dioctyl phthalate (117-81-7); hexahydrophthalic anhydrides (37226-48-5); himic anhydrides (2746-19-2); maleic anhydrides (108-31-6); methyltetrahydrophthalic anhydrides (26590-20-9); pyromellitic dianhydride (89-32-7); benzalkonium chloride (fumes) (8001-54-5); 1, 2-benzisothiazoline-3-one (fumes) (2634-33-5); bisulfite, SO_2:(7446-09-5); hydrobromic acid bromine; bromochlorodifluoromethane [halon 1211] (353-59-3); bromotrifluoromethane [halon 1301](75-63-8); calcium carbonate [chalk powder]; calcium oxide (1305-78-8); captafol (2425-06-1); chlorhexidine (55-56-1); aliphatic + cycloaliphatic diamine, (hardener) (2855-13-2) [isophorone diamine]; diazonium tetrafluoroborate (14239-22-6); dichlorvos [organophosphate] (62-73-7); diethanolamine (111-42-2); 2-dimethylaminoethanol [dimethyl ethanolamine] (108-01-0); ethylene oxide (75-21-8); freon-22; hexachlorophene (70-30-4); hydrazine (302-01-2); iridium salt; isocyanate prepolymers; 3-isocyanatomethyl-3,5,5-trimethylcyclohexyl isocyanate (4098-71-9); triglycidil isocyanurate (2451-62-1); triphenylmethane triisocyanate; isothiazolinone (55965-84-9); lauryl dimethyl benzyl ammonium chloride (139-07-1); metam sodium (144-54-7); methylmercaptan (74-93-1); monoethanolamine (141-43-5); ninhydrin (485-47-2); palladium (7440-05-3); ammonium persulfate; alcalic persulfate; sodium persulfate (7775-27-1); dipotassium peroxo-peroxodisulfate (7727-21-1); phosgene (75-44-5); polyethylene (9002-88-4); polymethyl-methacrylate (9011-14-7); sodium azide (powder dust) (26628-22-8); sulfathiazole (2-14-0); terpene (3-carene) (3466-78-9); tetrahydrothiophene (110-01-0); tetrachloroisophthalonitrile (fungicide); tetramethrin (7696-12-0); tributyl tin oxide; triethanolamine (102-71-6); tungsten carbide (11130-73-7); turpentine (8006-64-2); tylosin tartrate; uranium hexafluoride (7783-81-5); urea (fume)
		(57-13-6); urea formaldehyde foam (64869-57-4); phenol-formaldehyde resin (9003-35-4); zinc (fume) (7440-66-6); zinc chloride (fume) (7646-85-7); acid fluxes; acrylic acid; airbag content; bleaching agent (fumes); chlorofluorocarbons (degradation products); detergents; cutting oil; ECG ink; floor sealant (aromatic hydrocarbons); fumigating agent; furan-based binder; metal coating remover (coating removing chemical); metal oxide (fume); metal working fluids; perfume agents (research lab); pyrazolone; spray paint; tall oil; tear gas

CAS = Chemical abstracts service.

RCGP = Royal College of General Practitioners.

Low to moderate scientific evidence – provided by generally consistent findings in fewer, smaller or lower quality analytical studies, based on questionnaires or other inadequacies, i.e. "*[*]" – was found for 12 agents (phthalic anhydride, glutaraldehyde [glutaral], sulfur dioxide, cotton (dust, raw) CNT 750, potroom aluminum smelting, farming (various) or foundry), smoke (fires, pyrolysis products), pesticides (not specified), cleaning agents (not specified), ceramic production (dust), health care workers.

Limited or contradictory evidence – provided by only one analytical study or inconsistent findings in multiple scientific studies, i.e."*" – was identified for 39 agents, and after down-grading because of inadequate methodological aspects, i.e. "[*]" on three occasions. For the majority of agents, only non-analytical studies were reported for ≥ 5 cases, i.e. "(*)" or less than 5 cases, i.e. "-". When only non-analytical studies were available, the strength of evidence for the agent was raised if at least 5 cases were identified by the case reports/ case series or occupational disease statistics for which proof of irritant-induced OA or occupational COPD existed. The strength of evidence reached when only non-analytical studies were available ranged from "very limited or contradictory evidence" in 29 studies, i.e. "(*)", to "no scientific evidence" "-" 94 times. (see Tables 3 and 4 and Additional file 3: Table S2E of online supplement "Result").

The compiled assessment of the individual studies, along with their relevant clinical data and strength of evidence for irritant agents, professions or workplaces causing asthma or COPD, is presented as a summary list (see Additional file 3: Table S2E "Results" for the full information).

DISCUSSION

The main objective of this study was to give a comprehensive and evidence-based overview of the literature on irritative agents, professions or work-sites causing irritant-induced work-related asthma and occupational COPD. To our knowledge this study is the first attempts to document these respiratory disorders, along with their causative irritant agents in an evidence-based manner.

The 474 publications retrieved (see Table 3 and Additional file 3: Table S2E of online supplement "Results") in this work mainly refer to individual agents (n = 131), but also to mixed exposure(s) or multicomponent work-sites or professions (n = 57) where heterogeneous exposure to irritating substances is common, e.g. swine confinement, "construction work" or "farming", giving 188 different causes of irritant-induced OA and/or occupational COPD in total.

Strength and limitations

This work covers a broad range of causative agents of irritant-induced occupational asthma or COPD. We included various study designs.

A strength of our work is that we not only assessed the quality of single investigations but the strength of the body of evidence for each irritant agent.

The paradigm of "evidenced-based medicine" has been criticized by leading scientists [547–550]. Bias in the selection of information may be a problem for generalization of findings in single studies [551,552]. In spite of these limitations, alternative approaches to evaluation of the literature have not been generally accepted. Evaluation of the evidence depends on the domain, which means the factors to be considered in assessing the extent to which the study results are reliable or valid.

Kunz et al. [553] stressed the approach of grading scientific studies on basis of additional qualified data, i.e. dose response relationships. This latter was seen in 30/474 individual studies in this current work. Other studies were based on evidence by OR >2 or < 0.5 for irritant-induced OA and occupational COPD which was applied as an approach in 40/474 individual studies (see Additional file 3: Table S2E of online supplement "Results").

There are numerous procedural methods for rating the strength of scientific evidence. The AHRQ emphasized in 2002: "systems for grading the strength of a body of evidence are much less uniform than those for rating study quality" [554].

It is possible that not all relevant studies were found in our search of literature. Probably, some studies could not be found by the MeSH term raster applied. Relying solely on MeSH terms might be a problem in the identification of studies of irritant-induced OA or occupational COPD. We restricted the search to the MeSH fields in order to increase the specificity of the search. As for any electronic search strategy, an increase of specificity implies a decrease in sensitivity of the search.

For each single study, we took into consideration possible risks due to confounding, e.g. exposure to multiple agents and selection bias, e.g. healthy worker effect.

Basis and quality of data

Irritant-induced obstructive airways diseases cannot usually be diagnosed in one clinical visit and, instead, follow-up and/or detailed clinical investigations are necessary. The diagnostic "gold standard" for OA is SIC using a specific occupational agent in an exposure chamber. SIC is particularly indicated in the clinical setting where new causative substances with still unknown adverse respiratory sensitization potential are suspected. This "gold standard" is not applicable for large studies; so, it was used mainly in case series or reports.The evidence levels to confirm irritant-induced work-relaated asthma or occupational COPD for the listed irritant agents, professions or worksites (see Additional file 3: Table S2E of online supplement "Results") are frequently low with the major reasons being that high quality studies were missing and the quality of the available studies was low. Nevertheless, this knowledge is the best available and may help physicians to identify a suspected irritant agent as causative in irritant-induced work-related asthma and / or occupational COPD [555]. As also recently stressed by Quint et al. [555], "implementing an evidence-based identification and regulatory process for OA will help to ensure primary prevention of OA". In cases of low evidence level of an agent that does not exclude a causative role, caution should be exercised and a more detailed diagnostic testing of relevant exposure should be performed.

Occupational COPD, an underestimated category

We identified only 20 out of 474 publications that referred to occupational COPD, with most of them implicating inorganic or organic dust or fumes, such as cement dust, construction work and diesel exhaust, as the causative agents.

As an example, the mixed agent cement dust was investigated in 14 studies but only four studies documented cement dust as the causative agent in occupational COPD [111, 418, 419, 422] (see Table 3 and Additional file 3: Table S2E "Results"). The remaining 10 studies described irritant-induced OA cases [235, 423–426, 530] or identified significant asthma symptoms/ obstructive ventilation patterns without a clear diagnosis (5 studies: [178, 417, 420, 556]). It can be assumed that if it had been considered on the other 10 studies then occupational COPD caused by cement dust would have been frequently observed.

The population-attributable fraction for COPD associated with occupational exposure has been estimated between 9% and 31% [1, 64, 65]. However the true population-attributable risk due to occupational exposure is unclear [6, 557] as occupational COPD is rarely clinically diagnosed. Blanc et al. [558] recently published an ecological analysis using data from three large studies, comprising the Burden of Obstructive Lung Disease study [169], the Latin American Project for Investigation of Obstructive Lung Disease (PLATINO) and the European Community Respiratory Health Survey follow-up (ECHRS II), where occupational COPD was also not a primary goal. The original publications are mainly concerned with OA or asthma symptoms, but a history of pre-existing OA or RADS cannot be allowed to exclude occupational COPD [559]. Blanc et al. [558] stressed that the contribution of occupational exposure cannot be ignored, because "the association between adverse working conditions and COPD (...) carries significance as a global finding (...), alongside the (...) critical contribution of cigarette smoking to disease prevalence".

General acceptance of this statement does not exist [66, 559], although evidence for an association between individual exposure levels and COPD is accumulating in the latest literature [1, 6, 106, 506, 557, 560, 561].

Irritant-induced WRA – a broader definition

Irritant-induced OA includes three subcategories that predominantly differ according to the concentration of irritants in the workplace atmosphere. It can occur without a latency period, such as RADS, as was shown for 46 causative agents in our study, with the highest prevalence after spills of acids or tear gas (see Additional file 3: Table S2E of online supplement"Results"). Other agents, e.g. isocyanates or welding fumes, usually induce a slower onset of low dose irritant-induced asthma with a latency period and mostly without evidence of an IgE-mediated pathomechanism.

The ACCP also stated in its last Consensus Statement in 2008 [24] that cases who do not meet the stringent criteria of RADS [30] (e.g. where there is a lag of several days before the onset of symptoms or where there is no single massive exposure but rather repeated exposure over days and weeks) should be subsumed into a broader category of irritant-induced OA. As outlined in the section "Introduction" Brooks et al. [31] and later also others, e.g. Burge [27] suggested using the term "not so sudden onset of irritant-induced asthma" for those developing the disorder after such exposure within a period of 2 days to 4 months. In an extended definition corresponding to ours, Burge [27] he used the term "low dose irritant-induced OA" for those developing the disorder after relatively low repeated exposure for more than 4 months.

Bardana [562] and Vandenplas and Malo [563] questioned whether such rather low concentrations could actually cause irritant-induced OA. These different opinions about the pathogenetic role of chronic or recurrent exposure(s) to low concentrations of respiratory irritants seem to be due to inadequate considering of the increased susceptibility of a small group of workers. Occupational disease statistics do mostly neither contain such cases nor work-aggravated asthma cases so far.

Another critical issue is the frequent disregarding of work-aggravated asthma due to occupational agents by physicians.

Comparison to occupational guidelines or consensus statements – what is new?

In the current analysis, the focus has been on irritant agents causing irritant-induced occupational asthma and COPD. Both entities have been underestimated or even overlooked in the past. Occupational COPD has not been considered as a subgroup of COPD thus so far [559, 564]; and the definition of irritant-induced OA has been heterogenous at best [24, 552, 565, 566]. Furthermore, the guidelines dealing with respiratory disorders have not even considered causation by individual irritant agents, so far.

The ACCP published a Consensus Statement in 2008 [24] which focuses on the diagnosis and management of WRA after a latency period, i.e. due allergens and "sensitizers" with unknown pathomechanisms, effectively sidelining irritant-induced OA to RADS.

The Agency for Healthcare Research and Quality (AHRQ) in its the Evidence Report "Diagnosis and Management of WRA" [552] addressed the key question of the best diagnostic approach for a patient with suspected WRA. In respect of irritant-induced OA, they only considered RADS as a non-allergic asthma due to mainly low molecular weight compounds of unknown pathomechanism.

The Canadian Thoracic Society "Guidelines for OA" [567] was the first evidence-based guideline, although irritant-induced OA was limited to RADS. If criteria were not fulfilled then irritant-induced OA was discussed as a controversial diagnosis. The three evidence levels in the "Guidelines for OA" were based on quality of scientific evidence within analyzed studies [568]. Compared with the modified RCGP three-star grading (see Additional file 2: Table S2D in online supplement "Methodology"), the different levels are defined in a more general way, i.e. not considering the quantitative aspect if only studies with lower scientific evidence exist.

The evidence review and recommendations for OA by the BOHRF [3, 569] were designed to improve the prevention, identification and management of OA. This work mainly deals with asthma after a latency period and considers irritant-induced OA and RADS to be closely related entities. The difference in comparison with our analysis is obvious even though our evidence-based approach was closely related to the BOHRF guidelines and used the same grading systems.

In summary, the existing guidelines or statements mostly define irritant-induced OA as RADS. Work-aggravated asthma, and occupational COPD as a distinct entity, have not been considered in any guideline, although the latter is becoming recognized as such in more recent publications [557, 559,564].

This evidence-based approach is the first which focuses on especially irritative agents within the broader definition of irritant-induced OA and occupational COPD. For clarification, the grading systems were modified in accordance with BOHRF [3] when considering the extent and quality of the clinical investigations, with the goal of creating evidence levels for causative irritative agents as precisely as possible.

CONCLUDING REMARKS

OA is the most common chronic occupational lung disease in many industrialized countries [3]. COPD is the fourth leading cause of death worldwide with a significant portion of occupational cases [66]. The term occupational COPD does not officially exist. However, it has to be considered as a subcategory of COPD [559].

Our study shows that reliable, sensitive and specific methods are required in the diagnostic approach for confirming irritant-induced OA, work-aggravated asthma, or occupational COPD. The specific diagnostic work-up in a subject with such a suspected disorder depends on the individual clinical data and on the knowledge of asthma- or COPD-inducing agents in the workplace. On this basis, our review may help in diagnostics especially for agent exposures where we were able to relate irritant-induced work-related asthma or occupational COPD to a high evidence-based level (i.e. two stars according to the RCGP grading).

We have created a list representing the strength of evidence for irritating agents to be causative in irritant-induced work-related asthma or occupational COPD (see Additional file 3: Table S2E of online supplement "Results").

A low level or absence of evidence for many agents in causing irritant-induced work-related asthma or occupational COPD is sometimes due to contradictory findings in literature, but is mostly due to the absence of rigorous scientific studies, with many gaps remaining in the knowledge of a causative role for individual agents and conditions. Therefore, and because of rarely applied diagnostic approach in the clinical setting, our literature search and evaluation lead us to assume that irritant-induced respiratory disorders are considerably underreported in cross-sectional studies and occupational disease statistics.

Our list needs updating in the light of recent literature, in order to provide a realistic overview of agents and evidence level in their causation of irritant-induced work-related or occupational COPD.

The estimated high population-attributable risk in the range of 5–25% for occupational asthma and COPD from occupational exposure, indicates that more detailed and intensive research, as well as strategies designed to prevent these disorders, should receive high priority in the global efforts to reduce the burden of these diseases. This implies extended evidence-based diagnostic procedures that help to optimize primary and secondary prevention by the physicians dealing with occupational diseases.

Reduction of the exposure to noxious agents by lowering the permissible exposure limits is the best and favoured way for intervention. If this is not possible then other effective primary preventive measures, such as wearing adequate respiratory devices, are required [28, 570–574].

Finally, we would like to mention that the diagnosis of irritant-induced OA should be considered if:

- there has been exposure to high concentration of an irritative agent identified in this study and the development of asthma without a latency period (original definition of RADS) or
- there has been chronic or repeated exposures to moderate (in the TLV ranges) concentrations of an irritative agent identified in this review and the development of asthma with a latency period, but without evidence of an IgE-mediated pathomechanism and
- there is evidence that a highly susceptible subject (e.g. with pre-existing NSBHR) develops new onset asthma upon occupational exposure to an identified irritative agent even at concentrations below the TLV.

Work-aggravated asthma should be considered if:

- there have been any of the before-mentioned exposures and
- there is a temporally related significant worsening of a pre-existing asthma or of a concomitant non-occupational asthma.

The diagnosis of occupational COPD should be considered if:

- there has been exposure to an agent capable of causing occupational COPD, and
- not reversible chronic airway disease is demonstrated and
- there is a temporal relationship between the period of exposure (mostly cumulative exposures to identified irritants) and the development of COPD (acute WRA symptoms are frequently missing).

Occupational COPD has to be taken into consideration especially in non-smokers, i.e. when dominating non-occupational causes for COPD are obviously not present.

ENDNOTES

[a]Epidemiologic study design which is generally applied to test one or more specific hypotheses, typically whether an exposure is a risk factor for a disease [575].

AUTHORS' CONTRIBUTIONS

All authors made substantial contributions to the study. XB made the design of the study and the final interpretation of data. HV and PB did the detailed literature search, data extraction and analyses, and statistical analyses. XB and HV wrote the manuscript with input from PB. All authors approved the final version for submission.

REFERENCES

1. Balmes J, Becklake M, Blanc P, Henneberger P, Kreiss K, Mapp C, Milton D, Schwartz D, Toren K, Viegi G:**American Thoracic Society Statement: Occupational contribution to the burden of airway disease.** *Am J Respir Crit Care Med* 2003,**167**(5):787–797.

2. Bernstein IL, Chan-Yeung M, Malo JL, Bernstein DI: *Asthma in the workplace and related conditions*. 3rd edition. Taylor & Francis, New York; 2006.

3. Nicholson PJ, Cullinan P, Burge PS, Boyle C: *Occupational asthma: Prevention, identification & management: Systematic review & recommendations*. British Occupational Health Research Foundation, London; 2010.http://www.bohrf.org.uk/downloads/OccupationalAsthmaEvidenceReview-Mar2010.pdf

4. Arif AA, Whitehead LW, Delclos GL, Tortolero SR, Lee ES: **Prevalence and risk factors of work related asthma by industry among United States workers: data from the third**

national health and nutrition examination survey (1988–94). *Occup Environ Med* 2002,**59**(8):505–511.

5. Bakke PS, Baste V, Hanoa R, Gulsvik A: **Prevalence of obstructive lung disease in a general population: relation to occupational title and exposure to some airborne agents.** *Thorax* 1991,**46**(12):863–870.

6. Becklake MR: **Occupational exposures: evidence for a causal association with chronic obstructive pulmonary disease.** *Am Rev Respir Dis* 1989,**140**(3 Pt 2):S85-S91.

7. Becklake MR, Malo J-L, Chan-Yeung MY: **Epidemiological approaches in occupational asthma.** In *Asthma in the workplace*. 3rd edition. Edited by: Bernstein IL, Chan-Yeung M, Malo J-L, Bernstein DI. Taylor & Francis Group, New York, London; 2006:37–85.

8. Bergdahl IA, Toren K, Eriksson K, Hedlund U, Nilsson T, Flodin R, Jarvholm B: **Increased mortality in COPD among construction workers exposed to inorganic dust.** *Eur Respir J* 2004,**23**(3):402–406.

9. Blanc PD, Toren K: **How much adult asthma can be attributed to occupational factors?** *Am J Med*1999,**107**(6):580–587.

10. Henneberger PK, Derk SJ, Davis L, Tumpowsky C, Reilly MJ, Rosenman KD, Schill DP, Valiante D, Flattery J, Harrison R, *et al.*: **Work-related reactive airways dysfunction syndrome cases from surveillance in selected US states.** *J Occup Environ Med* 2003,**45**(4):360–368.

11. Karjalainen A, Kurppa K, Martikainen R, Klaukka T, Karjalainen J: **Work is related to a substantial portion of adult-onset asthma incidence in the Finnish population.** *Am J Respir Crit Care Med* 2001,**164**(4):565–568.

12. Kogevinas M, Anto JM, Sunyer J, Tobias A, Kromhout H, Burney P: **Occupational asthma in Europe and other industrialised areas: a population-based study. European Community Respiratory Health Survey Study Group.** *Lancet* 1999,**353**(9166):1750–1754.

13. Leuenberger P, Schindler C, Schwartz J, Ackermann-Liebrich U, Tara D, Perruchoud AP, Wuthrich B, Zellweger JP, Blaser K,

Bolognini G, *et al.*: **Occupational exposure to inhalative irritants and methacholine responsiveness.** *Scand J Work Environ Health* 2000,**26**(2):146–152.

14. Le Moual N, Kennedy SM, Kauffmann F: **Occupational exposures and asthma in 14,000 adults from the general population.** *Am J Epidemiol* 2004,**160**(11):1108–1116.

15. Medina-Ramon M, Zock JP, Kogevinas M, Sunyer J, Anto JM: **Asthma symptoms in women employed in domestic cleaning: a community based study.** *Thorax* 2003,**58**(11):950–954.

16. Meldrum M, Rawbone R, Curran AD, Fishwick D: **The role of occupation in the development of chronic obstructive pulmonary disease (COPD).** *Occup Environ Med* 2005,**62**(4):212–214.

17. Taylor AJ: **Respiratory irritants encountered at work.** *Thorax* 1996,**51**(5):541–545.

18. Petsonk EL: **Work-related asthma and implications for the general public.** *Environ Health Perspect*2002,**110**(Suppl 4):569–572.

19. Viegi G, Di Pede C: **Chronic obstructive lung diseases and occupational exposure.** *Curr Opin Allergy Clin Immunol* 2002,**2**(2):115–121.

20. WHO: **Reducing Risks, Promoting Healthy Life.** In *The World Health Report 2002*. WHO, Geneva; 2002.

21. Xu X, Christiani DC, Dockery DW, Wang L: **Exposure-response relationships between occupational exposures and chronic respiratory illness: a community-based study.** *Am Rev Respir Dis* 1992,**146**(2):413–418.

22. Toren K, Blanc PD: **Asthma caused by occupational exposures is common - a systematic analysis of estimates of the population-attributable fraction.** *BMC Pulm Med* 2009, **9:** 7.

23. Kogevinas M, Zock JP, Jarvis D, Kromhout H, Lillienberg L, Plana E, Radon K, Toren K, Alliksoo A, Benke G, *et al.*: **Exposure to substances in the workplace and new-onset asthma: an international prospective population-based study (ECRHS-II).** *Lancet* 2007,**370**(9584):336–341.

24. Tarlo SM, Balmes J, Balkissoon R, Beach J, Beckett W, Bernstein D: **Diagnosis and management of work-related asthma: American College of Chest Physicians Consensus Statement.** *Chest* 2008, **134:** 1–41.

25. Henneberger PK, Redlich CA, Callahan DB, Harber P, Lemiere C, Martin J, Tarlo SM, Vandenplas O, Toren K:**An official american thoracic society statement: work-exacerbated asthma.** *Am J Respir Crit Care Med*2011,**184**(3):368–378.

26. Lemiere C, Forget A, Dufour MH, Boulet LP, Blais L: **Characteristics and medical resource use of asthmatic subjects with and without work-related asthma.** *J Allergy Clin Immunol* 2007,**120**(6):1354–1359.

27. Burge SP, Moore VC, Robertson AS: **Sensitization and irritant-induced occupational asthma with latency are clinically indistinguishable.** *Occup Med (Lond)* 2012,**62**(2):129–133.

28. Baur X, Sigsgaard T, Aasen TB, Burge PS, Heederik D, Henneberger P, Maestrelli P, Rooyackers J, Schlunssen V, Vandenplas O, *et al.*: **Guidelines for the management of work-related asthma.** *Eur Respir J*2012,**39**(3):529–545.

29. Baur X, Aasen TB, Burge PS, Heederik D, Henneberger PK, Maestrelli P, Schlunssen V, Vandenplas O, Wilken D: **The management of work-related asthma guidelines: a broader perspective.** *Eur Respir Rev*2012,**21**(124):125–139.

30. Brooks SM, Weiss MA, Bernstein IL: **Reactive airways dysfunction syndrome (RADS). Persistent asthma syndrome after high level irritant exposures.** *Chest* 1985,**88**(3):376–384.

31. Brooks SM, Hammad Y, Richards I, Giovinco-Barbas J, Jenkins K: **The spectrum of irritant-induced asthma: sudden and not-so-sudden onset and the role of allergy.** *Chest* 1998,**113**(1):42–49.

32. Chan-Yeung M, Lam S, Kennedy SM, Frew AJ: **Persistent asthma after repeated exposure to high concentrations of gases in pulpmills.** *Am J Respir Crit Care Med* 1994,**149**(6):1676–1680.

33. Humerfelt S, Gulsvik A, Skjaerven R, Nilssen S, Kvale G, Sulheim O, Ramm E, Eilertsen E, Humerfelt SB:**Decline in FEV1 and**

airflow limitation related to occupational exposures in men of an urban community.*Eur Respir J* 1993,**6**(8):1095–1103.

34. Krzyzanowski M, Jedrychowski W, Wysocki M: **Factors associated with the change in ventilatory function and the development of chronic obstructive pulmonary disease in a 13-year follow-up of the Cracow Study. Risk of chronic obstructive pulmonary disease.** *Am Rev Respir Dis* 1986,**134**(5):1011–1019.

35. Tarlo SM, Broder I: **Irritant-induced occupational asthma.** *Chest* 1989,**96**(2):297–300.

36. Gautrin D, Bernstein IL, Brooks SM, Henneberger PK: **Reactive airways dysfunction syndrome and irritant-induced asthma.** In *Asthma in the workplace*. Edited by: Bernstein IL, Chan-Yeung M, Malo JL, Bernstein DI. Taylor & Francis Group, New York, London; 2006:581–629.

37. Kipen HM, Blume R, Hutt D: **Asthma experience in an occupational and environmental medicine clinic. Low-dose reactive airways dysfunction syndrome.** *J Occup Med* 1994,**36**(10):1133–1137.

38. Dykewicz MS: **Occupational asthma: current concepts in pathogenesis, diagnosis, and management.** *J Allergy Clin Immunol* 2009,**123**(3):519–528. quiz 529–530

39. Cormier Y, Coll B, Laviolette M, Boulet LP: **Reactive airways dysfunction syndrome (RADS) following exposure to toxic gases of a swine confinement building.** *Eur Respir J* 1996,**9**(5):1090–1091.

40. Dosman JA, Lawson JA, Kirychuk SP, Cornier Y, Biem J, Koehncke N: **Occupational asthma in newly employed workers in intensive swine confinement facilities.** *Eur Respir J* 2004, **24:** 698–702.

41. Reinisch F, Harrison RJ, Cussler S, Athanasoulis M, Balmes J, Blanc P, Cone J: **Physician reports of work-related asthma in California, 1993–1996.** *Am J Ind Med* 2001,**39**(1):72–83.

42. Liss GM, Tarlo SM, Doherty J, Purdham J, Greene J, McCaskell L, Kerr M: **Physician diagnosed asthma, respiratory symptoms, and associations with workplace tasks among radiographers in Ontario, Canada.***Occup Environ Med* 2003,**60**(4):254–261.

43. Smedley J, Coggon D: **Health surveillance for hospital employees exposed to respiratory sensitizers.** *Occup Med (Lond)* 1996,**46**(1):33–36.

44. Toren K, Jarvholm B, Brisman J, Hagberg S, Hermansson BA, Lillienberg L: **Adult-onset asthma and occupational exposures.** *Scand J Work Environ Health* 1999,**25**(5):430–435.

45. Makker HK, Ayres JG: **Work-related asthma in an aircraft engine mechanic.** *Respir Med* 1999,**93**(1):69–70.

46. Wade JF 3rd, Newman LS: **Diesel asthma. Reactive airways disease following overexposure to locomotive exhaust.** *J Occup Med* 1993,**35**(2):149–154.

47. Baur X: **Airborne allergens and irritants in the workplace.** In *Allergy and allergic diseases*. Edited by: Kay AB, Kaplan AP, Bousquet J, Holt PG. Blackwell Publishing, Boston; 2008:1017–1122.

48. Toren K, Balder B, Brisman J, Lindholm N, Lowhagen O, Palmqvist M, Tunsater A: **The risk of asthma in relation to occupational exposures: a case–control study from a Swedish city.** *Eur Respir J* 1999,**13**(3):496–501.

49. Hnizdo E, Sullivan PA, Bang KM, Wagner G: **Association between chronic obstructive pulmonary disease and employment by industry and occupation in the US population: a study of data from the Third National Health and Nutrition Examination Survey.** *Am J Epidemiol* 2002,**156**(8):738–746.

50. Oxman AD, Muir DC, Shannon HS, Stock SR, Hnizdo E, Lange HJ: **Occupational dust exposure and chronic obstructive pulmonary disease.** *A systematic overview of the evidence. Am Rev Respir Dis* 1993,**148**(1):38–48.

51. Marine WM, Gurr D, Jacobsen M: **Clinically important respiratory effects of dust exposure and smoking in British coal miners.** *Am Rev Respir Dis* 1988,**137**(1):106–112.

52. Kongerud J, Gronnesby JK, Magnus P: **Respiratory symptoms and lung function of aluminum potroom workers.** *Scand J Work Environ Health* 1990,**16**(4):270–277.

53. Andrasch RH, Bardana EJ Jr, Koster F, Pirofsky B: **Clinical and bronchial provocation studies in patients with meatwrappers' asthma.** *J Allergy Clin Immunol* 1976,**58**(2):291–298.

54. Carlsen KH, Anderson SD, Bjermer L, Bonini S, Brusasco V, Canonica W, Cummiskey J, Delgado L, Del Giacco SR, Drobnic F, *et al.*: **Exercise-induced asthma, respiratory and allergic disorders in elite athletes: epidemiology, mechanisms and diagnosis: Part I of the report from the Joint Task Force of the European Respiratory Society (ERS) and the European Academy of Allergy and Clinical Immunology (EAACI) in cooperation with GA2LEN.** *Allergy* 2008, **63:** 387–403.

55. Karjalainen A, Kurppa K, Virtanen S, Keskinen H, Nordman H: **Incidence of occupational asthma by occupation and industry in Finland.** *Am J Ind Med* 2000,**37**(5):451–458.

56. Hansson SO: **Critical effects and exposure limits.** *Risk Anal* 1997,**17**(2):227–236.

57. Latza U, Baur X: **Occupational obstructive airway diseases in Germany: Frequency and causes in an international comparison.** *Am J Ind Med* 2005,**48**(2):144–152.

58. Chan-Yeung M, Malo JL: **Aetiological agents in occupational asthma.** *Eur Respir J* 1994,**7**(2):346–371.

59. Fernandez-Nieto M, Quirce S, Sastre J: **Occupational asthma in industry.** *Allergol Immunopathol (Madr)*2006,**34**(5):212–223.

60. Baur X, Latza U: **Non-malignant occupational respiratory diseases in Germany in comparison with those of other countries.** *Int Arch Occup Environ Health* 2005,**78**(7):593–602.

61. Ameille J, Pauli G, Calastreng-Crinquand A, Vervloet D, Iwatsubo Y, Popin E, Bayeux-Dunglas MC, Kopferschmitt-Kubler MC: **Reported incidence of occupational asthma in France, 1996–99: the ONAP programme.** *Occup Environ Med* 2003,**60**(2):136–141.

62. Tarlo SM, Liss GM: **Occupational asthma: an approach to diagnosis and management.** *CMAJ*2003,**168**(7):867–871.

63. Pesci A, Majori M, Cuomo A, Borciani N, Bertacco S, Cacciani G, Gabrielli M: **Neutrophils infiltrating bronchial epithelium in**

chronic obstructive pulmonary disease. *Respir Med* 1998,**92**(6):863–870.

64. Hnizdo E, Glindmeyer HW, Petsonk EL, Enright P, Buist AS: **Case definitions for chronic obstructive pulmonary disease.** *Copd* 2006,**3**(2):95–100.

65. Trupin L, Earnest G, San Pedro M, Balmes JR, Eisner MD, Yelin E, Katz PP, Blanc PD: **The occupational burden of chronic obstructive pulmonary disease.** *Eur Respir J* 2003,**22**(3):462–469.

66. Balmes JR: **Occupational airways diseases from chronic low-level exposures to irritants.** *Clin Chest Med*2002,**23**(4):727–735.

67. Hendrick DJ: **Smoking, cadmium, and emphysema.** *Thorax* 2004,**59**(3):184–185.

68. Davison AG, Fayers PM, Taylor AJ, Venables KM, Darbyshire J, Pickering CA, Chettle DR, Franklin D, Guthrie CJ, Scott MC, *et al.*: **Cadmium fume inhalation and emphysema.** *Lancet* 1988,**1**(8587):663–667.

69. American Thoracic Society: **Standards for the diagnosis and care of patients with chronic obstructive pulmonary disease.** *Am J Respir Crit Care Med* 1995, **152:** 77–121.

70. Postma DS, Boezen HM: **Rationale for the Dutch hypothesis. Allergy and airway hyperresponsiveness as genetic factors and their interaction with environment in the development of asthma and COPD.** *Chest*2004,**126**(2 Suppl):96S-104S. discussion 159S-161S

71. Balmes JR: **Work-related COPD.** *Eur Respir J* 1994, **7:** 768–778.

72. Bourdin A, Serre I, Flamme H, Vic P, Neveu D, Aubas P, Godard P, Chanez P: **Can endobronchial biopsy analysis be recommended to discriminate between asthma and COPD in routine practice?** *Thorax*2004,**59**(6):488–493.

73. Hargreave FE, Parameswaran K: **Asthma, COPD and bronchitis are just components of airway disease.** *Eur Respir J* 2006,**28**(2):264–267.

74. Gannon PF, Burge PS: **The SHIELD scheme in the West Midlands Region, United Kingdom. Midland Thoracic Society Research Group.** *Br J Ind Med* 1993,**50**(9):791–796.

75. Eisner MD, Yelin EH, Katz PP, Lactao G, Iribarren C, Blanc PD: **Risk factors for work disability in severe adult asthma.** *Am J Med* 2006,**119**(10):884–891.

76. Toren K, Brisman J, Olin AC, Blanc PD: **Asthma on the job: work-related factors in new-onset asthma and in exacerbations of pre-existing asthma.** *Respir Med* 2000,**94**(6):529–535.

77. Brändli O, Schindler C, Leuenberger PH, Baur X, Degens P, Kunzli N, Keller R, Perruchoud AP: **Re-estimated equations for 5th percentiles of lung function variables.** *Thorax* 2000,**55**(2):173–174.

78. European Parliament, The Council of EU: **Regulation (EC) No 1272/2008 of the European Parliament and on the Council of 16 December 2008 on classification, labelling and packing of substances and mixtures, amending and repealing Directives 67/548/EEC and 1999/45/EC, and amending Regulations (EC) No 1907/2006.** *Off J EU* 2008, **L 353:** 1–1355. http://eur-lex.europa.eu/LexUriServ/LexUriServ.do?uri=OJ:L:2008:353:0001:0001:en:PDF

79. American Conference of Governmental Industrial Hygienists: *TLVs® and BEIs®. Threshold limit values for chemical substances and physical agents & biological exposure indices.* ACGIH, Cincinnaty; 2009.

80. Baur X: **Occupational agents with respiratory effects according to ACGIH 2009 and /or classified with the R42 phrase* (may cause sensitization by inhalation) and /or with the R37 phrase* (irritating to respiratory system) according to the European Union directives 67/548/EEC (1), 2001/59/EC (2), 2004/73/EC (3) or 2009/2/EC (4) (identical to ILO/CIS 2002.** 2009,http://www.ilo.org/public/english/protection/safework/cis/products/icsc/dtasht/riskphrs/index.htm) http://www.uke.de/institute/arbeitsmedizin/downloads/universitaetsprofessur-arbeitsmedizin/R42_und_R37A-EU09.pdf and www.eomsociety.org => knowledge center

81. Meyer JD, Holt DL, Chen Y, Cherry NM, McDonald JC: **SWORD '99: surveillance of work-related and occupational respiratory disease in the UK.** *Occup Med (Lond)* 2001,**51**(3):204–208.

82. Meyer JD, Holt DL, Cherry NM, McDonald JC: **SWORD '98: surveillance of work-related and occupational respiratory disease in the UK.** *Occup Med (Lond)* 1999,**49**(8):485–489.

83. Ross DJ, Keynes HL, McDonald JC: **SWORD '96: surveillance of work-related and occupational respiratory disease in the UK.** *Occup Med (Lond)* 1997,**47**(6):377–381.

84. Esterhuizen TM, Hnizdo E, Rees D: **Occurrence and causes of occupational asthma in South Africa–results from SORDSA's Occupational Asthma Registry, 1997–1999.** *S Afr Med J* 2001,**91**(6):509–513.

85. Buckley LA, Jiang XZ, James RA, Morgan KT, Barrow CS: **Respiratory tract lesions induced by sensory irritants at the RD50 concentration.** *Toxicol Appl Pharmacol* 1984,**74**(3):417–429.

86. Haupt B, Drechsel-Schlund C, Guldner K, Rogosky E, Plinske W, Butz M: *Dokumentation des Berufskrankheiten-Geschehens in Deutschland. BK-DOK 2005.* Deutsche Gesetzliche Unfallversicherung (DGUV), Sankt Augustin; 2007. http://www.dguv.de/inhalt/zahlen/documents/bk_dok_2005.pdf

87. Harbour R, Miller J: **A new system for grading recommendations in evidence based guidelines.** *BMJ*2001,**323**(7308):334–336.

88. The Royal College of General Practitioners RCGP: **The development and implementation of clinical guidelines.** In *Report of the Clinical Guidelines Working Group.* RCGP, London; 1995.

89. Alberts WM, do Pico GA: **Reactive airways dysfunction syndrome.** *Chest* 1996,**109**(6):1618–1626.

90. American Thoracic Society: **Guidelines for assessing and managing asthma risk at work, school, and recreation.** *Am J Respir Crit Care Med* 2004,**169**(7):873–881.

91. Baur X, Degens P, Weber K: **Occupational obstructive airway diseases in Germany.** *Am J Ind Med*1998,**33**(5):454–462.

92. van Kampen V, Merget R, Baur X: **Occupational airway sensitizers: an overview on the respective literature.** *Am J Ind Med* 2000,**38**(2):164–218.

93. **Webpage** http://www.hse.gov.uk

94. Malo JL: **Irritant-induced asthma and reactive airways dysfunction syndrome.** *Can Respir J* 1998,**5**(1):66–67.

95. Malo JL, Chan-Yeung M: **Asthma in the workplace: a Canadian contribution and perspective.** *Can Respir J*2007,**14**(7):407–413.

96. McDonald JC, Keynes HL, Meredith SK: **Reported incidence of occupational asthma in the United Kingdom, 1989–97.** *Occup Environ Med* 2000,**57**(12):823–829.

97. Sastre J, Vandenplas O, Park HS: **Pathogenesis of occupational asthma.** *Eur Respir J* 2003,**22**(2):364–373.

98. Joules H: **Asthma from sensitisation to chromium.** *Lancet* 1932, **2:** 182–183.

99. Boulet LP: **Increases in airway responsiveness following acute exposure to respiratory irritants. Reactive airway dysfunction syndrome or occupational asthma?** *Chest* 1988,**94**(3):476–481.

100. Simonsson BG, Sjoberg A, Rolf C, Haeger-Aronsen B: **Acute and long-term airway hyperreactivity in aluminium-salt exposed workers with nocturnal asthma.** *Eur J Respir Dis* 1985,**66**(2):105–118.

101. Matrat M, Laurence MF, Iwatsubo Y, Hubert C, Joly N, Legrand-Cattan K, L'Huillier JP, Villemain C, Pairon JC:**Reactive airways dysfunction syndrome caused by bromochlorodifluoromethane from fire extinguishers.***Occup Environ Med* 2004,**61**(8):712–714.

102. Cone JE, Wugofski L, Balmes JR, Das R, Bowler R, Alexeeff G, Shusterman D: **Persistent respiratory health effects after a metam sodium pesticide spill.** *Chest* 1994,**106**(2):500–508.

103. Valero AL, Bescos M, Amat P, Malet A: **Bronchial asthma caused by occupational sulfite exposure.** *Allergol Immunopathol (Madr)* 1993,**21**(6):221–224.

104. Cockcroft DW, Hoeppner VH, Dolovich J: **Occupational asthma caused by cedar urea formaldehyde particle board.** *Chest* 1982,**82**(1):49–53.

105. Hart JE, Laden F, Schenker MB, Garshick E: **Chronic obstructive pulmonary disease mortality in diesel-exposed railroad workers.** *Environ Health Perspect* 2006,**114**(7):1013–1017.

106. Hart JE, Laden F, Eisen EA, Smith TJ, Garshick E: **Chronic obstructive pulmonary disease mortality in railroad workers.** *Occup Environ Med* 2009,**66**(4):221–226.

107. Ulvestad B, Lund MB: **Increased risk of chronic obstructive pulmonary disease among tunnel construction workers.** *Tidsskr Nor Laegeforen* 2003,**123**(16):2292–2295.

108. Kern DG: **Outbreak of the reactive airways dysfunction syndrome after a spill of glacial acetic acid.** *Am Rev Respir Dis* 1991,**144**(5):1058–1064.

109. Gamble J, Jones W, Hancock J, Meckstroth RL: **Epidemiological-environmental study of lead acid battery workers. III. Chronic effects of sulfuric acid on the respiratory system and teeth.** *Environ Res*1984,**35**(1):30–52.

110. Bohadana AB, Massin N, Wild P, Berthiot G: **Airflow obstruction in chalkpowder and sugar workers.** *Int Arch Occup Environ Health* 1996,**68**(4):243–248.

111. Mwaiselage J, Bratveit M, Moen BE, Mashalla Y: **Respiratory symptoms and chronic obstructive pulmonary disease among cement factory workers.** *Scand J Work Environ Health* 2005,**31**(4):316–323.

112. Massin N, Hecht G, Ambroise D, Hery M, Toamain JP, Hubert G, Dorotte M, Bianchi B: **Respiratory symptoms and bronchial responsiveness among cleaning and disinfecting workers in the food industry.** *Occup Environ Med* 2007,**64**(2):75–81.

113. Sprince NL, Oliver LC, Eisen EA, Greene RE, Chamberlin RI: **Cobalt exposure and lung disease in tungsten carbide production. A cross-sectional study of current workers.** *Am Rev Respir Dis* 1988,**138**(5):1220–1226.

114. Woldeyohannes M, Bergevin Y, Mgeni AY, Theriault G: **Respiratory problems among cotton textile mill workers in Ethiopia.** *Br J Ind Med* 1991,**48**(2):110–115.

115. Rylander R, Bergstrom R: **Bronchial reactivity among cotton workers in relation to dust and endotoxin exposure.** *Ann Occup Hyg* 1993,**37**(1):57–63.

116. Rylander R, Haglind P, Lundholm M: **Endotoxin in cotton dust and respiratory function decrement among cotton workers in an experimental cardroom.** *Am Rev Respir Dis* 1985,**131**(2):209–213.

117. Chen R, Tunstall-Pedoe H, Tavendale R: **Environmental tobacco smoke and lung function in employees who never smoked: the Scottish MONICA study.** *Occup Environ Med* 2001,**58**(9):563–568.

118. Janson C, Chinn S, Jarvis D, Zock JP, Toren K, Burney P: **Effect of passive smoking on respiratory symptoms, bronchial responsiveness, lung function, and total serum IgE in the European Community Respiratory Health Survey: a cross-sectional study.** *Lancet* 2001,**358**(9299):2103–2109.

119. Hoppin JA, Umbach DM, London SJ, Alavanja MC, Sandler DP: **Diesel exhaust, solvents, and other occupational exposures as risk factors for wheeze among farmers.** *Am J Respir Crit Care Med*2004,**169**(12):1308–1313.

120. Melenka LS, Hessel PA, Yoshida K, Enarson DA: **Lung health in Alberta farmers.** *Int J Tuberc Lung Dis*1999,**3**(10):913–919.

121. Ahman M, Alexandersson R, Ekholm U, Bergstrom B, Dahlqvist M, Ulfvarson U: **Impeded lung function in moulders and coremakers handling furan resin sand.** *Int Arch Occup Environ Health* 1991,**63**(3):175–180.

122. Post W, Heederik D, Houba R: **Decline in lung function related to exposure and selection processes among workers in the grain processing and animal feed industry.** *Occup Environ Med* 1998,**55**(5):349–355.

123. Schwartz DA, Thorne PS, Yagla SJ, Burmeister LF, Olenchock SA, Watt JL, Quinn TJ: **The role of endotoxin in grain dust-induced lung disease.** *Am J Respir Crit Care Med* 1995,**152**(2):603–608.

124. Pahwa P, Senthilselvan A, McDuffie HH, Dosman JA: **Longitudinal estimates of pulmonary function decline in grain workers.** *Am J Respir Crit Care Med* 1994,**150**(3):656–662.

125. Wisnewski AV, Liu Q, Liu J, Redlich CA: **Glutathione protects human airway proteins and epithelial cells from isocyanates.** *Clin Exp Allergy* 2005,**35**(3):352–357.

126. Zock JP, Rodriguez-Trigo G, Pozo-Rodriguez F, Barbera JA, Bouso L, Torralba Y, Anto JM, Gomez FP, Fuster C, Verea H: **Prolonged respiratory symptoms in clean-up workers of the prestige oil spill.** *Am J Respir Crit Care Med* 2007,**176**(6):610–616.

127. Wieslander G, Norback D, Edling C: **Airway symptoms among house painters in relation to exposure to volatile organic compounds (VOCS)–a longitudinal study.** *Ann Occup Hyg* 1997,**41**(2):155–166.

128. Castro-Gutierrez N, McConnell R, Andersson K, Pacheco-Anton F, Hogstedt C: **Respiratory symptoms, spirometry and chronic occupational paraquat exposure.** *Scand J Work Environ Health* 1997,**23**(6):421–427.

129. Jones SM, Burks AW, Spencer HJ, Lensing S, Roberson PK, Gandy J, Helm RM: **Occupational asthma symptoms and respiratory function among aerial pesticide applicators.** *Am J Ind Med* 2003,**43**(4):407–417.

130. Hagmar L, Bellander T, Ranstam J, Skerfving S: **Piperazine-induced airway symptoms: exposure-response relationships and selection in an occupational setting.** *Am J Ind Med* 1984,**6**(5):347–357.

131. Calverley AE, Rees D, Dowdeswell RJ, Linnett PJ, Kielkowski D: **Platinum salt sensitivity in refinery workers: incidence and effects of smoking and exposure.** *Occup Environ Med* 1995,**52**(10):661–666.

132. Lorimer WV, Lilis R, Nicholson WJ, Anderson H, Fischbein A, Daum S, Rom W, Rice C, Selikoff IJ: **Clinical studies of styrene workers: initial findings.** *Environ Health Perspect* 1976, **17:** 171–181.

133. Prezant DJ, Weiden M, Banauch GI, McGuinness G, Rom WN, Aldrich TK, Kelly KJ: **Cough and bronchial responsiveness in**

firefighters at the World Trade Center site. *N Engl J Med* 2002,**347**(11):806–815.

134. Jaakkola MS, Leino T, Tammilehto L, Ylostalo P, Kuosma E, Alanko K: **Respiratory effects of exposure to methacrylates among dental assistants.** *Allergy* 2007,**62**(6):648–654.

135. Medina-Ramon M, Zock JP, Kogevinas M, Sunyer J, Torralba Y, Borrell A, Burgos F, Anto JM: **Asthma, chronic bronchitis, and exposure to irritant agents in occupational domestic cleaning: a nested case–control study.** *Occup Environ Med* 2005,**62**(9):598–606.

136. Sauni R, Oksa P, Huikko S, Roto P, Uitti J: **Increased risk of asthma among Finnish construction workers.***Occup Med (Lond)* 2003,**53**(8):527–531.

137. Jaakkola MS, Piipari R, Jaakkola N, Jaakkola JJ: **Environmental tobacco smoke and adult-onset asthma: a population-based incident case–control study.** *Am J Public Health* 2003,**93**(12):2055–2060.

138. Greer JR, Abbey DE, Burchette RJ: **Asthma related to occupational and ambient air pollutants in nonsmokers.** *J Occup Med* 1993,**35**(9):909–915.

139. Eisner MD, Balmes J, Katz PP, Trupin L, Yelin EH, Blanc PD: **Lifetime environmental tobacco smoke exposure and the risk of chronic obstructive pulmonary disease.** *Environ Health* 2005,**4**(1):7.

140. Fidan F, Cimrin AH, Ergor G, Sevinc C: **Airway disease risk from environmental tobacco smoke among coffeehouse workers in Turkey.** *Tob Control* 2004,**13**(2):161–166.

141. Radon K, Busching K, Heinrich J, Wichmann HE, Jorres RA, Magnussen H, Nowak D: **Passive smoking exposure: a risk factor for chronic bronchitis and asthma in adults?** *Chest* 2002,**122**(3):1086–1090.

142. Blanc PD, Ellbjar S, Janson C, Norback D, Norrman E, Plaschke P, Toren K: **Asthma-related work disability in Sweden. The impact of workplace exposures.** *Am J Respir Crit Care Med* 1999,**160**(6):2028–2033.

143. Flodin U, Jonsson P, Ziegler J, Axelson O: **An epidemiologic study of bronchial asthma and smoking.** *Epidemiology* 1995,**6**(5):503–505.

144. Eduard W, Douwes J, Omenaas E, Heederik D: **Do farming exposures cause or prevent asthma? Results from a study of adult Norwegian farmers.** *Thorax* 2004,**59**(5):381–386.

145. Hoppin JA, Umbach DM, London SJ, Alavanja MC, Sandler DP: **Animal production and wheeze in the Agricultural Health Study: interactions with atopy, asthma, and smoking.** *Occup Environ Med*2003,**60**(8):e3.

146. Melbostad E, Eduard W, Magnus P: **Determinants of asthma in a farming population.** *Scand J Work Environ Health* 1998,**24**(4):262–269.

147. Mastrangelo G, Tartari M, Fedeli U, Fadda E, Saia B: **Ascertaining the risk of chronic obstructive pulmonary disease in relation to occupation using a case–control design.** *Occup Med (Lond)* 2003,**53**(3):165–172.

148. Mastrangelo G, Paruzzolo P, Mapp C: **Asthma due to isocyanates: a mail survey in a 1% sample of furniture workers in the Veneto region, Italy.** *Med Lav* 1995,**86**(6):503–510.

149. Meredith SK, Bugler J, Clark RL: **Isocyanate exposure and occupational asthma: a case-referent study.** *Occup Environ Med* 2000,**57**(12):830–836.

150. Jacobs JH, Spaan S, van Rooy GB, Meliefste C, Zaat VA, Rooyackers JM, Heederik D: **Exposure to trichloramine and respiratory symptoms in indoor swimming pool workers.** *Eur Respir J* 2007,**29**(4):690–698.

151. Schenker MB, Stoecklin M, Lee K, Lupercio R, Zeballos RJ, Enright P, Hennessy T, Beckett LA: **Pulmonary function and exercise-associated changes with chronic low-level paraquat exposure.** *Am J Respir Crit Care Med* 2004,**170**(7):773–779.

152. Senthilselvan A, McDuffie HH, Dosman JA: **Association of asthma with use of pesticides. Results of a cross-sectional survey of farmers.** *Am Rev Respir Dis* 1992,**146**(4):884–887.

153. Danuser B, Weber C, Kunzli N, Schindler C, Nowak D: **Respiratory symptoms in Swiss farmers: an epidemiological study of risk factors.** *Am J Ind Med* 2001,**39**(4):410–418.

154. Andersson E, Nilsson T, Persson B, Wingren G, Toren K: **Mortality from asthma and cancer among sulfite mill workers.** *Scand J Work Environ Health* 1998,**24**(1):12–17.

155. Preller L, Heederik D, Boleij JS, Vogelzang PF, Tielen MJ: **Lung function and chronic respiratory symptoms of pig farmers: focus on exposure to endotoxins and ammonia and use of disinfectants.** *Occup Environ Med*1995,**52**(10):654–660.

156. Wheeler K, McKelvey W, Thorpe L, Perrin M, Cone J, Kass D, Farfel M, Thomas P, Brackbill R: **Asthma diagnosed after 11 September 2001 among rescue and recovery workers: findings from the World Trade Center Health Registry.** *Environ Health Perspect* 2007,**115**(11):1584–1590.

157. Rajan KG, Davies BH: **Reversible airways obstruction and interstitial pneumonitis due to acetic acid.** *Br J Ind Med* 1989,**46**(1):67–68.

158. Kivity S, Fireman E, Lerman Y: **Late asthmatic response to inhaled glacial acetic acid.** *Thorax*1994,**49**(7):727–728.

159. Moore VC, Manney S, Vellore AD, Burge PS: **Occupational asthma to gel flux containing dodecanedioic acid.***Allergy* 2009,**64**(7):1099–1100.

160. Musk AW, Peach S, Ryan G: **Occupational asthma in a mineral analysis laboratory.** *Br J Ind Med*1988,**45**(6):381–386.

161. Deschamps D, Soler P, Rosenberg N, Baud F, Gervais P: **Persistent asthma after inhalation of a mixture of sodium hypochlorite and hydrochloric acid.** *Chest* 1994,**105**(6):1895–1896.

162. Promisloff RA, Lenchner GS, Phan A, Cichelli AV: **Reactive airway dysfunction syndrome in three police officers following a roadside chemical spill.** *Chest* 1990,**98**(4):928–929.

163. Franzblau A, Sahakian N: **Asthma following household exposure to hydrofluoric acid.** *Am J Ind Med*2003,**44**(3):321–324.

164. el-Sadik YM, Osman HA, el-Gazzar RM: **Exposure to sulfuric acid in manufacture of storage batteries.** *J Occup Med* 1972,**14**(3):224–226.

165. Chatkin JM, Tarlo SM, Liss G, Banks D, Broder I: **The outcome of asthma related to workplace irritant exposures: a comparison of irritant-induced asthma and irritant aggravation of asthma.** *Chest*1999,**116**(6):1780–1785.

166. Savonius B, Keskinen H, Tuppurainen M, Kanerva L: **Occupational respiratory disease caused by acrylates.***Clin Exp Allergy* 1993,**23**(5):416–424.

167. Yacoub MR, Lemiere C, Malo JL: **Asthma caused by cyanoacrylate used in a leisure activity.** *J Allergy Clin Immunol* 2005,**116**(2):462.

168. Nakazawa T: **Occupational asthma due to alkyl cyanoacrylate.** *J Occup Med* 1990,**32**(8):709–710.

169. Lozewicz S, Davison AG, Hopkirk A, Burge PS, Boldy DA, Riordan JF, McGivern DV, Platts BW, Davies D, Newman Taylor AJ: **Occupational asthma due to methyl methacrylate and cyanoacrylates.** *Thorax*1985,**40**(11):836–839.

170. Chan CC, Cheong TH, Lee HS, Wang YT, Poh SC: **Case of occupational asthma due to glue containing cyanoacrylate.** *Ann Acad Med Singapore* 1994,**23**(5):731–733.

171. Quirce S, Baeza ML, Tornero P, Blasco A, Barranco R, Sastre J: **Occupational asthma caused by exposure to cyanoacrylate.** *Allergy* 2001,**56**(5):446–449.

172. Kopp SK, McKay RT, Moller DR, Cassedy K, Brooks SM: **Asthma and rhinitis due to ethylcyanoacrylate instant glue.** *Ann Intern Med* 1985,**102**(5):613–615.

173. Weytjens K, Cartier A, Lemiere C, Malo JL: **Occupational asthma to diacrylate.** *Allergy* 1999,**54**(3):289–290.

174. Pickering CA, Bainbridge D, Birtwistle IH, Griffiths DL: **Occupational asthma due to methyl methacrylate in an orthopaedic theatre sister.** *Br Med J (Clin Res Ed)* 1986,**292**(6532):1362–1363.

175. Savonius B, Keskinen H, Tuppurainen M, Kanerva L: **Occupational asthma caused by ethanolamines.** *Allergy*1994,**49**(10):877–881.

176. Pepys J, Pickering CA: **Asthma due to inhaled chemical fumes–amino-ethyl ethanolamine in aluminium soldering flux.** *Clin Allergy* 1972,**2**(2):197–204.

177. Hnizdo E, Sylvain D, Lewis DM, Pechter E, Kreiss K: **New-onset asthma associated with exposure to 3-amino-5-mercapto-1,2,4-triazole.** *J Occup Environ Med* 2004,**46**(12):1246–1252.

178. Ali BA, Ahmed HO, Ballal SG, Albar AA: **Pulmonary function of workers exposed to ammonia: a study in the Eastern Province of Saudi Arabia.** *Int J Occup Environ Health* 2001,**7**(1):19–22.

179. Bernstein DI: *Guidelines for the diagnosis and evaluation of occupational immunologic lung disease*. Mosby Co, St. Louis; 1989.

180. de la Hoz RE, Schlueter DP, Rom WN: **Chronic lung disease secondary to ammonia inhalation injury: a report on three cases.** *Am J Ind Med* 1996,**29**(2):209–214.

181. Leduc D, Gris P, Lheureux P, Gevenois PA, De Vuyst P, Yernault JC: **Acute and long term respiratory damage following inhalation of ammonia.** *Thorax* 1992,**47**(9):755–757.

182. Flury KE, Dines DE, Rodarte JR, Rodgers R: **Airway obstruction due to inhalation of ammonia.** *Mayo Clin Proc* 1983,**58**(6):389–393.

183. Weir DC, Robertson AS, Jones S, Burge PS: **Occupational asthma due to soft corrosive soldering fluxes containing zinc chloride and ammonium chloride.** *Thorax* 1989,**44**(3):220–223.

184. Gelfand HH: **Respiratory Allergy Due to Chemical Compounds Encountered in the Rubber, Lacquer, Shellac, and Beauty Culture Industries.** *J Allergy Clin Immunol* 1963, **34:** 374–381.

185. Greene SA, Freedman S: **Asthma due to inhaled chemical agents–amprolium hydrochloride.** *Clin Allergy*1976,**6**(2):105–108.

186. Baur X, Czuppon AB, Rauluk I, Zimmermann FB, Schmitt B, Egen-Korthaus M, Tenkhoff N, Degens PO: **A clinical and**

immunological study on 92 workers occupationally exposed to anhydrides. *Int Arch Occup Environ Health* 1995,**67**(6):395–403.

187. Drexler H, Weber A, Letzel S, Kraus G, Schaller KH, Lenhert G: **Detection and clinical relevance of a type I allergy with occupational exposure to hexahydrophthalic anhydride and methyltetrahydrophthalic anhydride.** *Int Arch Occup Environ Health* 1994,**65**(5):279–283.

188. Cipolla C, Belisario A, Sassi C, Auletti G, Nobile M, Raffi GB: **Occupational asthma caused by dioctyl-phthalate in a bottle cap production worker.** *Med Lav* 1999,**90**(3):513–518.

189. Chee CB, Lee HS, Cheong TH, Wang YT, Poh SC: **Occupational asthma due to hexahydrophthalic anhydride: a case report.** *Br J Ind Med* 1991,**48**(9):643–645.

190. Rosenman KD, Bernstein DI, O'Leary K, Gallagher JS, D'Souza L, Bernstein IL: **Occupational asthma caused by himic anhydride.** *Scand J Work Environ Health* 1987,**13**(2):150–154.

191. Lee HS, Wang YT, Cheong TH, Tan KT, Chee BE, Narendran K: **Occupational asthma due to maleic anhydride.** *Br J Ind Med* 1991,**48**(4):283–285.

192. Graneek BJ, Durham SR, Newman Taylor AJ: **Late asthmatic reactions and changes in histamine responsiveness provoked by occupational agents.** *Bull Eur Physiopathol Respir* 1987,**23**(6):577–581.

193. Nielsen J, Welinder H, Skerfving S: **Allergic airway disease caused by methyl tetrahydrophthalic anhydride in epoxy resin.** *Scand J Work Environ Health* 1989,**15**(2):154–155.

194. Nielsen J, Welinder H, Schutz A, Skerfving S: **Specific serum antibodies against phthalic anhydride in occupationally exposed subjects.** *J Allergy Clin Immunol* 1988,**82**(1):126–133.

195. Wernfors M, Nielsen J, Schutz A, Skerfving S: **Phthalic anhydride-induced occupational asthma.** *Int Arch Allergy Appl Immunol* 1986,**79**(1):77–82.

196. Frans A, Pahulycz C: **Transient syndrome of acute irritation of the bronchi induced by single and massive inhalation of phthalic anhydride.** *Rev Pneumol Clin* 1993,**49**(5):247–251.

197. Fawcett IW, Taylor AJ, Pepys J: **Asthma due to inhaled chemical agents–epoxy resin systems containing phthalic acid anhydride, trimellitic acid anhydride and triethylene tetramine.** *Clin Allergy* 1977,**7**(1):1–14.

198. Maccia CA, Bernstein IL, Emmett EA, Brooks SM: **In vitro demonstration of specific IgE in phthalic anhydride hypersensitivity.** *Am Rev Respir Dis* 1976,**113**(5):701–704.

199. Meadway J: **Asthma and atopy in workers with an epoxy adhesive.** *Br J Dis Chest* 1980,**74**(2):149–154.

200. Venables KM, Topping MD, Howe W, Luczynska CM, Hawkins R, Taylor AJ: **Interaction of smoking and atopy in producing specific IgE antibody against a hapten protein conjugate.** *Br Med J (Clin Res Ed)*1985,**290**(6463):201–204.

201. Venables KM, Topping MD, Nunn AJ, Howe W, Newman Taylor AJ: **Immunologic and functional consequences of chemical (tetrachlorophthalic anhydride)-induced asthma after four years of avoidance of exposure.** *J Allergy Clin Immunol* 1987,**80**(2):212–218.

202. Howe W, Venables KM, Topping MD, Dally MB, Hawkins R, Law JS, Taylor AJ: **Tetrachlorophthalic anhydride asthma: evidence for specific IgE antibody.** *J Allergy Clin Immunol* 1983,**71**(1 Pt 1):5–11.

203. Schlueter DP, Banaszak EF, Fink JN, Barboriak J: **Occupational asthma due to tetrachlorophthalic anhydride.** *J Occup Med* 1978,**20**(3):183–188.

204. Grammer LC, Shaughnessy MA, Kenamore BD: **Clinical and immunologic outcome of 42 individuals with trimellitic anhydride-induced immunologic lung disease after transfer to low exposure.** *Allergy Asthma Proc* 2000,**21**(6):355–359.

205. Grammer LC, Shaughnessy MA, Kenamore BD, Yarnold PR: **A clinical and immunologic study to assess risk of TMA-induced lung disease as related to exposure.** *J Occup Environ Med* 1999,**41**(12):1048–1051.

206. Grammer L, Shaughnessy M, Kenamore B: **Utility of antibody in identifying individuals who have or will develop anhydride-induced respiratory disease.** *Chest* 1998,**114**(4):1199–1202.

207. Zeiss CR, Patterson R, Pruzansky JJ, Miller MM, Rosenberg M, Levitz D: **Trimellitic anhydride-induced airway syndromes: clinical and immunologic studies.** *J Allergy Clin Immunol* 1977,**60**(2):96–103.

208. Kanerva L, Keskinen H, Autio P, Estlander T, Tuppurainen M, Jolanki R: **Occupational respiratory and skin sensitization caused by polyfunctional aziridine hardener.** *Clin Exp Allergy* 1995,**25**(5):432–439.

209. Leffler CT, Milton DK: **Occupational asthma and contact dermatitis in a spray painter after introduction of an aziridine cross-linker.** *Environ Health Perspect* 1999,**107**(7):599–601.

210. Slovak AJ: **Occupational asthma caused by a plastics blowing agent, azodicarbonamide.** *Thorax*1981,**36**(12):906–909.

211. Kim CW, Cho JH, Leem JH, Ryu JS, Lee HL, Hong YC: **Occupational asthma due to azodicarbonamide.** *Yonsei Med J* 2004,**45**(2):325–329.

212. Normand JC, Grange F, Hernandez C, Ganay A, Davezies P, Bergeret A, Prost G: **Occupational asthma after exposure to azodicarbonamide: report of four cases.** *Br J Ind Med* 1989,**46**(1):60–62.

213. Valentino M, Comai M: **Occupational asthma caused by azodicarbonamide: clinical case.** *G Ital Med Lav*1985,**7**(2–3):97–99.

214. Malo JL, Pineau L, Cartier A: **Occupational asthma due to azobisformamide.** *Clin Allergy* 1985,**15**(3):261–264.

215. Purohit A, Kopferschmitt-Kubler MC, Moreau C, Popin E, Blaumeiser M, Pauli G: **Quaternary ammonium compounds and occupational asthma.** *Int Arch Occup Environ Health* 2000,**73**(6):423–427.

216. Bernstein JA, Stauder T, Bernstein DI, Bernstein IL: **A combined respiratory and cutaneous hypersensitivity syndrome induced by work exposure to quaternary amines.** *J Allergy Clin Immunol* 1994,**94**(2 Pt 1):257–259.

217. Moscato G, Omodeo P, Dellabianca A, Colli MC, Pugliese F, Locatelli C, Scibilia J: **Occupational asthma and rhinitis caused**

by 1,2-benzisothiazolin-3-one in a chemical worker. *Occup Med (Lond)* 1997,**47**(4):249–251.

218. Burns MJ, Linden CH: **Another hot tub hazard. Toxicity secondary to bromine and hydrobromic acid exposure.** *Chest* 1997,**111**(3):816–819.

219. de la Hoz RE: **Reactive airways dysfunction syndrome following exposure to a fluorocarbon.** *Eur Respir J*1999,**13**(5):1192–1194.

220. Chan OY, Poh SC, Lee HS, Tan KT, Kwok SF: **Respiratory function in cadmium battery workers–a follow-up study.** *Ann Acad Med Singapore* 1988,**17**(2):283–287.

221. Leduc D, de Francquen P, Jacobovitz D, Vandeweyer R, Lauwerys R, De Vuyst P: **Association of cadmium exposure with rapidly progressive emphysema in a smoker.** *Thorax* 1993,**48**(5):570–571.

222. De Silva PE, Donnan MB: **Chronic cadmium poisoning in a pigment manufacturing plant.** *Br J Ind Med*1981,**38**(1):76–86.

223. Royce S, Wald P, Sheppard D, Balmes J: **Occupational asthma in a pesticides manufacturing worker.** *Chest*1993,**103**(1):295–296.

224. Neghab M, Mohraz MH, Hassanzadeh J: **Symptoms of respiratory disease and lung functional impairment associated with occupational inhalation exposure to carbon black dust.** *J Occup Health* 2011,**53**(6):432–438.

225. Bourne MS, Flindt ML, Walker JM: **Asthma due to industrial use of chloramine.** *BMJ* 1979,**2**(6181):10–12.

226. Feinberg SM, Watrons RM: **Atopy to simple chemical compounds-sulfonechloramides.** *J Allergy Clin Immunol* 1945,**16**(5):209–220.

227. Kujala VM, Reijula KE, Ruotsalainen EM, Heikkinen K: **Occupational asthma due to chloramine-T solution.***Respir Med* 1995,**89**(10):693–695.

228. Jouannique V, Pillière F, Pouillard D, Gervais P, Rousselin X, Efthymiou ML: **Asthme professionel à la chloramine T.** In *A propos de deux observations*. Société De Médicine et d'Hygiène du Travail, Paris; 1992:654–657.

229. Dijkman JH, Vooren PH, Kramps JA: **Occupational asthma due to inhalation of chloramine-T. I. Clinical observations and inhalation-provocation studies.** *Int Arch Allergy Appl Immunol* 1981,**64**(4):422–427.

230. Waclawski ER, McAlpine LG, Thomson NC: **Occupational asthma in nurses caused by chlorhexidine and alcohol aerosols.** *Bmj* 1989,**298**(6678):929–930.

231. Glindmeyer HW, Lefante JJ, Freyder LM, Friedman M, Weill H, Jones RN: **Relationship of asthma to irritant gas exposures in pulp and paper mills.** *Respir Med* 2003,**97**(5):541–548.

232. Andersson E, Olin AC, Hagberg S, Nilsson R, Nilsson T, Toren K: **Adult-onset asthma and wheeze among irritant-exposed bleachery workers.** *Am J Ind Med* 2003,**43**(5):532–538.

233. Gautrin D, Leroyer C, L'Archeveque J, Dufour JG, Girard D, Malo JL: **Cross-sectional assessment of workers with repeated exposure to chlorine over a three year period.** *Eur Respir J* 1995,**8**(12):2046–2054.

234. Bherer L, Cushman R, Courteau JP, Quevillon M, Cote G, Bourbeau J, L'Archeveque J, Cartier A, Malo JL:**Survey of construction workers repeatedly exposed to chlorine over a three to six month period in a pulpmill: II. Follow up of affected workers by questionnaire, spirometry, and assessment of bronchial responsiveness 18 to 24 months after exposure ended.** *Occup Environ Med* 1994,**51**(4):225–228.

235. Leroyer C, Dewitte JD, Bassanets A, Boutoux M, Daniel C, Clavier J: **Occupational asthma due to chromium.***Respiration* 1998,**65**(5):403–405.

236. Chester EH, Gillespie DG, Krause FD: **The prevalence of chronic obstructive pulmonary disease in chlorine gas workers.** *Am Rev Respir Dis* 1969,**99**(3):365–373.

237. Ferris BG Jr, Burgess WA, Worcester J: **Prevalence of chronic respiratory disease in a pulp mill and a paper mill in the United States.** *Br J Ind Med* 1967,**24**(1):26–37.

238. Lemiere C, Malo JL, Boutet M: **Reactive airways dysfunction syndrome due to chlorine: sequential bronchial biopsies and functional assessment.** *Eur Respir J* 1997,**10**(1):241–244.

239. Schonhofer B, Voshaar T, Kohler D: **Long-term lung sequelae following accidental chlorine gas exposure.***Respiration* 1996,**63**(3):155–159.

240. Olaguibel JM, Basomba A: **Occupational asthma induced by chromium salts.** *Allergol Immunopathol (Madr)*1989,**17**(3):133–136.

241. Onizuka R, Tanabe K, Nakayama Y, Fukuchi T, Nakata K, Hiki T: **A case of chrome asthma induced by exposure to the stone cutter dust.** *Arerugi* 2006,**55**(12):1556–1561.

242. Fernandez-Nieto M, Quirce S, Carnes J, Sastre J: **Occupational asthma due to chromium and nickel salts.** *Int Arch Occup Environ Health* 2006,**79**(6):483–486.

243. Sastre J, Fernandez-Nieto M, Maranon F, Fernandez-Caldas E, Pelta R, Quirce S: **Allergenic cross-reactivity between nickel and chromium salts in electroplating-induced asthma.** *J Allergy Clin Immunol*2001,**108**(4):650–651.

244. Nagasaka Y, Nakano N, Tohda Y, Nakajima S: **Persistent reactive airway dysfunction syndrome after exposure to chromate].** *Nihon Kyobu Shikkan Gakkai Zasshi* 1995,**33**(7):759–764.

245. Park HS, Yu HJ, Jung KS: **Occupational asthma caused by chromium.** *Clin Exp Allergy* 1994,**24**(7):676–681.

246. Novey HS, Habib M, Wells ID: **Asthma and IgE antibodies induced by chromium and nickel salts.** *J Allergy Clin Immunol* 1983,**72**(4):407–412.

247. Bernstein IL, Merget R: **Metals.** In *Asthma in the workplace*. 3rd edition. Edited by: Bernstein IL, Chan-Yeung M, Malo JL, Bernstein DI. Taylor & Francis, New York; 2006:525–554.

248. Linna A, Oksa P, Palmroos P, Roto P, Laippala P, Uitti J: **Respiratory health of cobalt production workers.** *Am J Ind Med* 2003,**44**(2):124–132.

249. Kusaka Y, Iki M, Kumagai S, Goto S: **Epidemiological study of hard metal asthma.** *Occup Environ Med*1996,**53**(3):188–193.

250. Kennedy SM, Chan-Yeung M, Marion S, Lea J, Teschke K: **Maintenance of stellite and tungsten carbide saw tips:**

respiratory health and exposure-response evaluations. *Occup Environ Med* 1995,**52**(3):185–191.

251. Roto P: **Asthma, symptoms of chronic bronchitis and ventilatory capacity among cobalt and zinc production workers.** *Scand J Work Environ Health* 1980,**6**(Suppl 1):1–49.

252. Pisati G, Zedda S: **Outcome of occupational asthma due to cobalt hypersensitivity.** *Sci Total Environ*1994,**150**(1–3):167–171.

253. Shirakawa T, Kusaka Y, Fujimura N, Goto S, Kato M, Heki S, Morimoto K: **Occupational asthma from cobalt sensitivity in workers exposed to hard metal dust.** *Chest* 1989,**95**(1):29–37.

254. Shirakawa T, Kusaka Y, Fujimura N, Goto S, Morimoto K: **The existence of specific antibodies to cobalt in hard metal asthma.** *Clin Allergy* 1988,**18**(5):451–460.

255. Shirakawa T, Kusaka Y, Fujimura N, Kato M, Heki S, Morimoto K: **Hard metal asthma: cross immunological and respiratory reactivity between cobalt and nickel?** *Thorax* 1990,**45**(4):267–271.

256. Gheysens B, Auwerx J, Van den Eeckhout A, Demedts M: **Cobalt-induced bronchial asthma in diamond polishers.** *Chest* 1985,**88**(5):740–744.

257. Baik JJ, Yoon YB, Park HS: **Cobalt-induced occupational asthma associated with systemic illness.** *J Korean Med Sci* 1995,**10**(3):200–204.

258. Wilk-Rivard E, Szeinuk J: **Occupational asthma with paroxysmal atrial fibrillation in a diamond polisher.***Environ Health Perspect* 2001,**109**(12):1303–1306.

259. Krakowiak A, Dudek W, Tarkowski M, Swiderska-Kielbik S, Niescierenko E, Palczynski C: **Occupational asthma caused by cobalt chloride in a diamond polisher after cessation of occupational exposure: a case report.***Int J Occup Med Environ Health* 2005,**18**(2):151–158.

260. Sargent EV, Brubaker RE, Mitchell CA: **Respiratory effects of occupational exposure to an epoxy resin system.** *Arch Environ Health* 1976,**31**(5):236–240.

261. Aleva RM, Aalbers R, Koeter GH, De Monchy JG: **Occupational asthma caused by a hardener containing an aliphatic and a cycloaliphatic diamine.** *Am Rev Respir Dis* 1992,**145**(5):1217–1218.

262. Luczynska CM, Hutchcroft BJ, Harrison MA, Dornan JD, Topping MD: **Occupational asthma and specific IgE to a diazonium salt intermediate used in the polymer industry.** *J Allergy Clin Immunol* 1990,**85**(6):1076–1082.

263. Graham VA, Coe MJ, Davies RJ: **Occupational asthma after exposure to a diazonium salt.** *Thorax*1981,**36**(12):950–951.

264. Emad A, Rezaian GR: **The diversity of the effects of sulfur mustard gas inhalation on respiratory system 10 years after a single, heavy exposure: analysis of 197 cases.** *Chest* 1997,**112**(3):734–738.

265. Deschamps D, Questel F, Baud FJ, Gervais P, Dally S: **Persistent asthma after acute inhalation of organophosphate insecticide.** *Lancet* 1994,**344**(8938):1712.

266. Barthel E: **Irritative and allergic effects of pesticide aerosols on the respiratory tract and problems of their evaluation.** *Z Gesamte Hyg* 1983,**29**(11):678–681.

267. Piipari R, Tuppurainen M, Tuomi T, Mantyla L, Henriks-Eckerman ML, Keskinen H, Nordman H:**Diethanolamine-induced occupational asthma, a case report.** *Clin Exp Allergy* 1998,**28**(3):358–362.

268. Gadon ME, Melius JM, McDonald GJ, Orgel D: **New-onset asthma after exposure to the steam system additive 2-diethylaminoethanol.** *A descriptive study. J Occup Med* 1994,**36**(6):623–626.

269. Vallieres M, Cockcroft DW, Taylor DM, Dolovich J, Hargreave FE: **Dimethyl ethanolamine-induced asthma.***Am Rev Respir Dis* 1977,**115**(5):867–871.

270. Cockcroft DW, Cotton DJ, Mink JT: **Nonspecific bronchial hyperreactivity after exposure to Western Red Cedar.** *Am Rev Respir Dis* 1979,**119**(3):505–510.

271. Conrad E, Lo W, de Boisblanc BP, Shellito JE: **Reactive airways dysfunction syndrome after exposure to dinitrogen tetroxide.** *South Med J* 1998,**91**(4):338–341.

272. Aldrich FD, Stange AW, Geesaman RE: **Smoking and ethylene diamine sensitization in an industrial population.** *J Occup Med* 1987,**29**(4):311–314.

273. Casas X, Badorrey I, Monso E, Morera J: **Occupational asthma due to amines].** *Arch Bronconeumol*2002,**38**(2):93–94.

274. Hagmar L, Bellander T, Bergoo B, Simonsson BG: **Piperazine-induced occupational asthma.** *J Occup Med*1982,**24**(3):193–197.

275. Lam S, Chan-Yeung M: **Ethylenediamine-induced asthma.** *Am Rev Respir Dis* 1980,**121**(1):151–155.

276. Deschamps D, Rosenberg N, Soler P, Maillard G, Fournier E, Salson D, Gervais P: **Persistent asthma after accidental exposure to ethylene oxide.** *Br J Ind Med* 1992,**49**(7):523–525.

277. Dugue P, Faraut C, Figueredo M, Bettendorf A, Salvadori JM: **Occupational asthma provoked by ethylene oxide in a nurse.** *Presse Med* 1991,**20**(30):1455.

278. Nunn AJ, Craigen AA, Darbyshire JH, Venables KM, Newman Taylor AJ: **Six year follow up of lung function in men occupationally exposed to formaldehyde.** *Br J Ind Med* 1990,**47**(11):747–752.

279. Nordman H, Keskinen H, Tuppurainen M: **Formaldehyde asthma–rare or overlooked?** *J Allergy Clin Immunol*1985,**75**(1 Pt 1):91–99.

280. Burge PS, Harries MG, Lam WK, O'Brien IM, Patchett PA: **Occupational asthma due to formaldehyde.** *Thorax*1985,**40**(4):255–260.

281. Piipari R, Keskinen H: **Agents causing occupational asthma in Finland in 1986–2002: cow epithelium bypassed by moulds from moisture-damaged buildings.** *Clin Exp Allergy* 2005,**35**(12):1632–1637.

282. Kim CW, Song JS, Ahn YS, Park SH, Park JW, Noh JH, Hong CS: **Occupational asthma due to formaldehyde.** *Yonsei Med J* 2001,**42**(4):440–445.

283. Lemiere C, Desjardins A, Cloutier Y, Drolet D, Perrault G, Cartier A, Malo JL: **Occupational asthma due to formaldehyde resin dust with and without reaction to formaldehyde gas.** *Eur Respir J* 1995,**8**(5):861–865.

284. Porter JA: **Letter: Acute respiratory distress following formalin inhalation.** *Lancet* 1975,**2**(7935):603–604.

285. Hendrick DJ, Lane DJ: **Formalin asthma in hospital staff.** *BMJ* 1975,**1**(5958):607–608.

286. Sjogren B, Gunnare S, Sandler H: **Inhalation of decomposed chlorodifluoromethane (freon-22) and myocardial infarction.** *Scand J Work Environ Health* 2002,**28**(3):205–207.

287. Malo JL, Gagnon G, Cartier A: **Occupational asthma due to heated freon.** *Thorax* 1984,**39**(8):628–629.

288. Vyas A, Pickering CA, Oldham LA, Francis HC, Fletcher AM, Merrett T, Niven RM: **Survey of symptoms, respiratory function, and immunology and their relation to glutaraldehyde and other occupational exposures among endoscopy nursing staff.** *Occup Environ Med* 2000,**57**(11):752–759.

289. Curran AD, Burge PS, Wiley K: **Clinical and immunologic evaluation of workers exposed to glutaraldehyde.** *Allergy* 1996,**51**(11):826–832.

290. Gannon PF, Bright P, Campbell M, O'Hickey SP, Burge PS: **Occupational asthma due to glutaraldehyde and formaldehyde in endoscopy and x ray departments.** *Thorax* 1995,**50**(2):156–159.

291. Jachuck SJ, Bound CL, Steel J, Blain PG: **Occupational hazard in hospital staff exposed to 2 per cent glutaraldehyde in an endoscopy unit.** *J Soc Occup Med* 1989,**39**(2):69–71.

292. Ong TH, Tan KL, Lee HS, Eng P: **A case report of occupational asthma due to gluteraldehyde exposure.** *Ann Acad Med Singapore* 2004,**33**(2):275–278.

293. Cullinan P, Hayes J, Cannon J, Madan I, Heap D, Taylor AN: **Occupational asthma in radiographers.** *Lancet*1992,**340**(8833):1477.

294. Corrado OJ, Osman J, Davies RJ: **Asthma and rhinitis after exposure to glutaraldehyde in endoscopy units.***Hum Toxicol* 1986,**5**(5):325–328.

295. Nagy L, Orosz M: **Occupational asthma due to hexachlorophene.** *Thorax* 1984,**39**(8):630–631.

296. Merget R, Topcu M, Friese K, Vormberg R, Fuchs T, Raulf-Heimsoth M, Breitstadt R: **A cross-sectional study of workers in the chemical industry with occupational exposure to hexamethylenetetramine.** *Int Arch Occup Environ Health* 1999,**72**(8):533–538.

297. Gamble JF, McMichael AJ, Williams T, Battigelli M: **Respiratory function and symptoms: an environmental-epidemiological study of rubber workers exposed to a phenolformaldehyde type resin.** *Am Ind Hyg Assoc J* 1976,**37**(9):499–513.

298. Bergman A, Svedberg U, Nilsson E: **Contact urticaria with anaphylactic reactions caused by occupational exposure to iridium salt.** *Contact Dermatitis* 1995,**32**(1):14–17.

299. Baur X, Dewair M, Fruhmann G: **Detection of immunologically sensitized isocyanate workers by RAST and intracutaneous skin tests.** *J Allergy Clin Immunol* 1984,**73**(5 Pt 1):610–618.

300. Piirila PL, Nordman H, Keskinen HM, Luukkonen R, Salo SP, Tuomi TO, Tuppurainen M: **Long-term follow-up of hexamethylene diisocyanate-, diphenylmethane diisocyanate-, and toluene diisocyanate-induced asthma.** *Am J Respir Crit Care Med* 2000,**162**(2 Pt 1):516–522.

301. O'Brien IM, Harries MG, Burge PS, Pepys J: **Toluene di-isocyanate-induced asthma. I. Reactions to TDI, MDI, HDI and histamine.** *Clin Allergy* 1979,**9**(1):1–6.

302. Cartier A, Grammer L, Malo JL, Lagier F, Ghezzo H, Harris K, Patterson R: **Specific serum antibodies against isocyanates: association with occupational asthma.** *J Allergy Clin Immunol* 1989,**84**(4 Pt 1):507–514.

303. Tarlo SM, Liss GM, Yeung KS: **Changes in rates and severity of compensation claims for asthma due to diisocyanates: a possible effect of medical surveillance measures.** *Occup Environ Med* 2002,**59**(1):58–62.

304. Bernstein DI, Korbee L, Stauder T, Bernstein JA, Scinto J, Herd ZL, Bernstein IL: **The low prevalence of occupational asthma and antibody-dependent sensitization to diphenylmethane diisocyanate in a plant engineered for minimal exposure to diisocyanates.** *J Allergy Clin Immunol* 1993,**92**(3):387–396.

305. Liss GM, Bernstein DI, Moller DR, Gallagher JS, Stephenson RL, Bernstein IL: **Pulmonary and immunologic evaluation of foundry workers exposed to methylene diphenyldiisocyanate (MDI).** *J Allergy Clin Immunol*1988,**82**(1):55–61.

306. Woellner RC, Hall S, Greaves I, Schoenwetter WF: **Epidemic of asthma in a wood products plant using methylene diphenyl diisocyanate.** *Am J Ind Med* 1997,**31**(1):56–63.

307. Mapp CE, Corona PC, De Marzo N, Fabbri L: **Persistent asthma due to isocyanates. A follow-up study of subjects with occupational asthma due to toluene diisocyanate (TDI).** *Am Rev Respir Dis* 1988,**137**(6):1326–1329.

308. Zammit-Tabona M, Sherkin M, Kijek K, Chan H, Chan-Yeung M: **Asthma caused by diphenylmethane diisocyanate in foundry workers. Clinical, bronchial provocation, and immunologic studies.** *Am Rev Respir Dis* 1983,**128**(2):226–230.

309. Perfetti L, Brame B, Ferrari M, Moscato G: **Occupational asthma (OA) with sensitization to diphenylmethane diisocyanate (MDI) presenting at the onset like a reactive airways dysfunction syndrome (RADS).** *Am J Ind Med* 2003,**44**(3):325–328.

310. Hur GY, Koh DH, Choi GS, Park HJ, Choi SJ, Ye YM, Kim KS, Park HS: **Clinical and immunologic findings of methylene diphenyl diisocyanate-induced occupational asthma in a car upholstery factory.** *Clin Exp Allergy* 2008,**38**(4):586–593.

311. Vandenplas O, Malo JL, Dugas M, Cartier A, Desjardins A, Levesque J, Shaughnessy MA, Grammer LC:**Hypersensitivity pneumonitis-like reaction among workers exposed to**

diphenylmethane [correction to piphenylmethane] diisocyanate (MDI). *Am Rev Respir Dis* 1993,**147**(2):338–346.

312. Lemiere C, Malo JL, Boulet LP, Boutet M: **Reactive airways dysfunction syndrome induced by exposure to a mixture containing isocyanate: functional and histopathologic behaviour.** *Allergy* 1996,**51**(4):262–265.

313. Belin L, Hjortsberg U, Wass U: **Life-threatening pulmonary reaction to car paint containing a prepolymerized isocyanate.** *Scand J Work Environ Health* 1981,**7**(4):310–311.

314. Vandenplas O, Cartier A, Lesage J, Cloutier Y, Perreault G, Grammer LC, Shaughnessy MA, Malo JL:**Prepolymers of hexamethylene diisocyanate as a cause of occupational asthma.** *J Allergy Clin Immunol*1993,**91**(4):850–861.

315. Clarke CW, Aldons PM: **Isophorone diisocyanate induced respiratory disease (IPDI).** *Aust N Z J Med*1981,**11**(3):290–292.

316. Kamat SR, Patel MH, Pradhan PV, Taskar SP, Vaidya PR, Kolhatkar VP, Gopalani JP, Chandarana JP, Dalal N, Naik M: **Sequential respiratory, psychologic, and immunologic studies in relation to methyl isocyanate exposure over two years with model development.** *Environ Health Perspect* 1992, **97:** 241–253.

317. Baur X, Chen Z, Marczynski B: **Respiratory diseases caused by occupational exposure to 1,5-naphthalene-diisocyanate (NDI): Results of workplace-related challenge tests and antibody analyses.** *Am J Ind Med*2001,**39**(4):369–372.

318. Vijayan VK, Sankaran K: **Relationship between lung inflammation, changes in lung function and severity of exposure in victims of the Bhopal tragedy.** *Eur Respir J* 1996,**9**(10):1977–1982.

319. Kamat SR, Mahashur AA, Tiwari AK, Potdar PV, Gaur M, Kolhatkar VP, Vaidya P, Parmar D, Rupwate R, Chatterjee TS, *et al.*: **Early observations on pulmonary changes and clinical morbidity due to the isocyanate gas leak at Bhopal.** *J Postgrad Med* 1985,**31**(2):63–72.

320. Harries MG, Burge PS, Samson M, Taylor AJ, Pepys J: **Isocyanate asthma: respiratory symptoms due to 1,5-naphthylene di-isocyanate.** *Thorax* 1979,**34**(6):762–766.

321. Mehta PS, Mehta AS, Mehta SJ, Makhijani AB: **Bhopal tragedy's health effects.** *A review of methyl isocyanate toxicity. JAMA* 1990,**264**(21):2781–2787.

322. Alexandersson R, Gustafsson P, Hedenstierna G, Rosen G: **Exposure to naphthalene-diisocyanate in a rubber plant: symptoms and lung function.** *Arch Environ Health* 1986,**41**(2):85–89.

323. Baur X, Wieners D, Marczynski B: **Late asthmatic reaction caused by naphthylene-1,5 diisocyanate.** *Scand J Work Environ Health* 2000,**26**(1):78–80.

324. Fuortes LJ, Kiken S, Makowsky M: **An outbreak of naphthalene di-isocyanate-induced asthma in a plastics factory.** *Arch Environ Health* 1995,**50**(5):337–340.

325. Seguin P, Allard A, Cartier A, Malo JL: **Prevalence of occupational asthma in spray painters exposed to several types of isocyanates, including polymethylene polyphenylisocyanate.** *J Occup Med* 1987,**29**(4):340–344.

326. Butcher BT, Jones RN, O'Neil CE, Glindmeyer HW, Diem JE, Dharmarajan V, Weill H, Salvaggio JE: **Longitudinal study of workers employed in the manufacture of toluene-diisocyanate.** *Am Rev Respir Dis*1977,**116**(3):411–421.

327. Ott MG, Klees JE, Poche SL: **Respiratory health surveillance in a toluene di-isocyanate production unit, 1967–97: clinical observations and lung function analyses.** *Occup Environ Med* 2000,**57**(1):43–52.

328. Omae K, Higashi T, Nakadate T, Tsugane S, Nakaza M, Sakurai H: **Four-year follow-up of effects of toluene diisocyanate exposure on the respiratory system in polyurethane foam manufacturing workers. II. Four-year changes in the effects on the respiratory system.** *Int Arch Occup Environ Health* 1992,**63**(8):565–569.

329. Marabini A, Brugnami G, Curradi F, Severini C, Siracusa A: **The response to a specific bronchial provocation test and the**

evolution of occupational asthma. A longitudinal study in subjects with toluene diisocyanate-induced asthma. *Med Lav* 1994,**85**(2):134–141.

330. Karol MH, Tollerud DJ, Campbell TP, Fabbri L, Maestrelli P, Saetta M, Mapp CE: **Predictive value of airways hyperresponsiveness and circulating IgE for identifying types of responses to toluene diisocyanate inhalation challenge.** *Am J Respir Crit Care Med* 1994,**149**(3 Pt 1):611–615.

331. Paggiaro P, Bacci E, Paoletti P, Bernard P, Dente FL, Marchetti G, Talini D, Menconi GF, Giuntini C:**Bronchoalveolar lavage and morphology of the airways after cessation of exposure in asthmatic subjects sensitized to toluene diisocyanate.** *Chest* 1990,**98**(3):536–542.

332. Axford AT, McKerrow CB, Jones AP, Le Quesne PM: **Accidental exposure to isocyanate fumes in a group of firemen.** *Br J Ind Med* 1976,**33**(2):65–71.

333. Vandenplas O, Cartier A, Lesage J, Perrault G, Grammer LC, Malo JL: **Occupational asthma caused by a prepolymer but not the monomer of toluene diisocyanate (TDI).** *J Allergy Clin Immunol* 1992,**89**(6):1183–1188.

334. Luo JC, Nelsen KG, Fischbein A: **Persistent reactive airway dysfunction syndrome after exposure to toluene diisocyanate.** *Br J Ind Med* 1990,**47**(4):239–241.

335. Piirila P, Estlander T, Hytonen M, Keskinen H, Tupasela O, Tuppurainen M: **Rhinitis caused by ninhydrin develops into occupational asthma.** *Eur Respir J* 1997,**10**(8):1918–1921.

336. Buick JB, Todd GR: **Concomitant alveolitis and asthma following exposure to triphenylmethane triisocyanate.** *Occup Med (Lond)* 1997,**47**(8):504–506.

337. Bourke SJ, Convery RP, Stenton SC, Malcolm RM, Hendrick DJ: **Occupational asthma in an isothiazolinone manufacturing plant.** *Thorax* 1997,**52**(8):746–748.

338. Burge PS, Richardson MN: **Occupational asthma due to indirect exposure to lauryl dimethyl benzyl ammonium chloride used in a floor cleaner.** *Thorax* 1994,**49**(8):842–843.

339. Belin L, Wass U, Audunsson G, Mathiasson L: **Amines: possible causative agents in the development of bronchial hyperreactivity in workers manufacturing polyurethanes from isocyanates.** *Br J Ind Med*1983,**40**(3):251–257.

340. Estlander T, Kanerva L, Tupasela O, Keskinen H, Jolanki R: **Immediate and delayed allergy to nickel with contact urticaria, rhinitis, asthma and contact dermatitis.** *Clin Exp Allergy* 1993,**23**(4):306–310.

341. Block GT, Yeung M: **Asthma induced by nickel.** *JAMA* 1982,**247**(11):1600–1602.

342. Malo JL, Cartier A, Doepner M, Nieboer E, Evans S, Dolovich J: **Occupational asthma caused by nickel sulfate.***J Allergy Clin Immunol* 1982,**69**(1 Pt 1):55–59.

343. McConnell LH, Fink JN, Schlueter DP, Schmidt MG Jr: **Asthma caused by nickel sensitivity.** *Ann Intern Med*1973,**78**(6):888–890.

344. Thickett KM, McCoach JS, Gerber JM, Sadhra S, Burge PS: **Occupational asthma caused by chloramines in indoor swimming-pool air.** *Eur Respir J* 2002,**19**(5):827–832.

345. Mehta AJ, Henneberger PK, Toren K, Olin AC: **Airflow limitation and changes in pulmonary function among bleachery workers.** *Eur Respir J* 2005,**26**(1):133–139.

346. Daenen M, Rogiers P, Van de Walle C, Rochette F, Demedts M, Nemery B: **Occupational asthma caused by palladium.** *Eur Respir J* 1999,**13**(1):213–216.

347. Silberman DE, Sorrell AH: **Allergy in fur workers with special reference to paraphenylenediamine.** *J Allergy*1959,**30**(1):11–18.

348. Moscato G, Pignatti P, Yacoub MR, Romano C, Spezia S, Perfetti L: **Occupational asthma and occupational rhinitis in hairdressers.** *Chest* 2005,**128**(5):3590–3598.

349. Pankow W, Hein H, Bittner K, Wichert P: **Persulfate asthma in hairdressers.** *Pneumologie* 1989,**43**(3):173–175.

350. Harth V, Raulf-Heimsoth M, Bruning T, Merget R: **Isolated late asthmatic reaction after exposure to ammonium persulfate in a hairdresser.** *Contact Dermatitis* 2006,**54**(1):62–63.

351. Wrbitzky R, Drexler H, Letzel S: **Early reaction type allergies and diseases of the respiratory passages in employees from persulphate production.** *Int Arch Occup Environ Health* 1995,**67**(6):413–417.

352. Therond M, Geraut C, Dupas D, Gayoux C: *Pathology des persulfates alcalins*. 1989, 837–838.

353. Parra FM, Igea JM, Quirce S, Ferrando MC, Martin JA, Losada E: **Occupational asthma in a hairdresser caused by persulphate salts.** *Allergy* 1992,**47**(6):656–660.

354. Munoz X, Cruz MJ, Orriols R, Bravo C, Espuga M, Morell F: **Occupational asthma due to persulfate salts: diagnosis and follow-up.** *Chest* 2003,**123**(6):2124–2129.

355. Blainey AD, Ollier S, Cundell D, Smith RE, Davies RJ: **Occupational asthma in a hairdressing salon.** *Thorax*1986,**41**(1):42–50.

356. Gamboa PM, de la Cuesta CG, Garcia BE, Castillo JG, Oehling A: **Late asthmatic reaction in a hairdresser, due to the inhalation of ammonium persulphate salts.** *Allergol Immunopathol (Madr)* 1989,**17**(2):109–111.

357. Baur X, Fruhmann G, von Liebe V: **Occupational asthma and dermatitis after exposure to dusts of persulfate salts in two industrial workers (author's transl).** *Respiration* 1979,**38**(3):144–150.

358. Kammermeyer JK, Mathews KP: **Hypersensitivity to phenylglycine acid chloride.** *J Allergy Clin Immunol*1973,**52**(2):73–84.

359. Wyatt JP, Allister CA: **Occupational phosgene poisoning: a case report and review.** *J Accid Emerg Med*1995,**12**(3):212–213.

360. Merget R, Kulzer R, Dierkes-Globisch A, Breitstadt R, Gebler A, Kniffka A, Artelt S, Koenig HP, Alt F, Vormberg R, *et al.*: **Exposure-effect relationship of platinum salt allergy in a catalyst production plant: conclusions from a 5-year

prospective cohort study. *J Allergy Clin Immunol* 2000,**105**(2 Pt 1):364–370.

361. Hnizdo E, Esterhuizen TM, Rees D, Lalloo UG: **Occupational asthma as identified by the Surveillance of Work-related and Occupational Respiratory Diseases programme in South Africa.** *Clin Exp Allergy*2001,**31**(1):32–39.

362. Merget R, Reineke M, Rueckmann A, Bergmann EM, Schultze-Werninghaus G: **Nonspecific and specific bronchial responsiveness in occupational asthma caused by platinum salts after allergen avoidance.** *Am J Respir Crit Care Med* 1994,**150**(4):1146–1149.

363. Baker DB, Gann PH, Brooks SM, Gallagher J, Bernstein IL: **Cross-sectional study of platinum salts sensitization among precious metals refinery workers.** *Am J Ind Med* 1990,**18**(6):653–664.

364. Venables KM, Dally MB, Nunn AJ, Stevens JF, Stephens R, Farrer N, Hunter JV, Stewart M, Hughes EG, Newman Taylor AJ: **Smoking and occupational allergy in workers in a platinum refinery.** *Bmj* 1989,**299**(6705):939–942.

365. Pickering CA: **Inhalation tests with chemical allergens: complex salts of platinum.** *Proc R Soc Med*1972,**65**(3):272–274.

366. Stenton SC, Kelly CA, Walters EH, Hendrick DJ: **Occupational asthma due to a repair process for polyethylene-coated electrical cables.** *J Soc Occup Med* 1989,**39**(1):33–34.

367. Gannon PF, Burge PS, Benfield GF: **Occupational asthma due to polyethylene shrink wrapping (paper wrapper's asthma).** *Thorax* 1992,**47**(9):759.

368. Skerfving S, Akesson B, Simonsson BG: **"Meat wrappers' asthma" caused by thermal degradation products of polyethylene.** *Lancet* 1980,**1**(8161):211.

369. Kennes B, Garcia-Herreros P, Dierckx P: **Asthma from plexiglas powders.** *Clin Allergy* 1981,**11**(1):49–54.

370. Atis S, Tutluoglu B, Levent E, Ozturk C, Tunaci A, Sahin K, Saral A, Oktay I, Kanik A, Nemery B: **The respiratory effects of occupational polypropylene flock exposure.** *Eur Respir J* 2005,**25**(1):110–117.

371. Malo JL, Cartier A, Boulet LP, L'Archeveque J, Saint-Denis F, Bherer L, Courteau JP: **Bronchial hyperresponsiveness can improve while spirometry plateaus two to three years after repeated exposure to chlorine causing respiratory symptoms.** *Am J Respir Crit Care Med* 1994,**150**(4):1142–1145.

372. Tuomainen A, Stark H, Seuri M, Hirvonen MR, Linnainmaa M, Sieppi A, Tukiainen H: **Experimental PVC material challenge in subjects with occupational PVC exposure.** *Environ Health Perspect* 2006,**114**(9):1409–1413.

373. Lee HS, Ng TP, Ng YL, Phoon WH: **Diurnal variation in peak expiratory flow rate among polyvinylchloride compounding workers.** *Br J Ind Med* 1991,**48**(4):275–278.

374. Wegman DH, Smith TJ, Eisen EA, Greaves IA, Fine LJ, Chelton CS: **Respiratory effects of work in retail food stores. I. Methodology and exposure assignments.** *Scand J Work Environ Health* 1987,**13**(3):203–208.

375. Baser ME, Tockman MS, Kennedy TP: **Pulmonary function and respiratory symptoms in polyvinylchloride fabrication workers.** *Am Rev Respir Dis* 1985,**131**(2):203–208.

376. Munoz X, Cruz MJ, Albanell M, Morell F: **Occupational asthma in food packers.** *Arch Bronconeumol*2003,**39**(7):324–326.

377. Lee HS, Yap J, Wang YT, Lee CS, Tan KT, Poh SC: **Occupational asthma due to unheated polyvinylchloride resin dust.** *Br J Ind Med* 1989,**46**(11):820–822.

378. Sokol WN, Aelony Y, Beall GN: **Meat-wrapper's asthma. A new syndrome?** *Jama* 1973,**226**(6):639–641.

379. Bright P, Burge PS, O'Hickey SP, Gannon PF, Robertson AS, Boran A: **Occupational asthma due to chrome and nickel electroplating.** *Thorax* 1997,**52**(1):28–32.

380. Hjortsberg U, Nise G, Orbaek P, Soes-Petersen U, Arborelius M Jr: **Bronchial asthma due to exposure to potassium aluminumtetrafluoride.** *Scand J Work Environ Health* 1986,**12**(3):223.

381. Burge PS, Edge G, Hawkins R, White V, Taylor AJ: **Occupational asthma in a factory making flux-cored solder containing colophony.** *Thorax* 1981,**36**(11):828–834.

382. Burge PS, Harries MG, O'Brien I, Pepys J: **Bronchial provocation studies in workers exposed to the fumes of electronic soldering fluxes.** *Clin Allergy* 1980,**10**(2):137–149.

383. Maestrelli P, Alessandri MV, Dal Vecchio L, Bartolucci GB, Cocheo V: **Occupational asthma due to colophony.***Med Lav* 1985,**76**(5):371–378.

384. Weiss JS: **Reactive airway dysfunction syndrome due to sodium azide inhalation.** *Int Arch Occup Environ Health* 1996,**68**(6):469–471.

385. Stenton SC, Dennis JH, Walters EH, Hendrick DJ: **Asthmagenic properties of a newly developed detergent ingredient: sodium iso-nonanoyl oxybenzene sulphonate.** *Br J Ind Med* 1990,**47**(6):405–410.

386. Hendrick DJ, Connolly MJ, Stenton SC, Bird AG, Winterton IS, Walters EH: **Occupational asthma due to sodium iso-nonanoyl oxybenzene sulphonate, a newly developed detergent ingredient.** *Thorax*1988,**43**(6):501–502.

387. Ferguson H, Thomas KE, Ollier S, Davies RJ: **Bronchial provocation testing of sodium iso-nonanoyl oxybenzene sulphonate.** *Hum Exp Toxicol* 1990,**9**(2):83–89.

388. Merget R, Korn M: **Metabisulphite-induced occupational asthma in a radiographer.** *Eur Respir J*2005,**25**(2):386–388.

389. Madsen J, Sherson D, Kjoller H, Hansen I, Rasmussen K: **Occupational asthma caused by sodium disulphite in Norwegian lobster fishing.** *Occup Environ Med* 2004,**61**(10):873–874.

390. Malo JL, Cartier A, Desjardins A: **Occupational asthma caused by dry metabisulphite.** *Thorax*1995,**50**(5):585–586. discussion 589

391. Pougnet R, Lodde B, Lucas D, Jegaden D, Bell S, Dewitte JD: **A case of occupational asthma from metabisulphite in a fisherman.** *Int Marit Health* 2010,**62**(3):180–184.

392. Steiner M, Scaife A, Semple S, Hulks G, Ayres JG: **Sodium metabisulphite induced airways disease in the fishing and fish-processing industry.** *Occup Med (Lond)* 2008,**58**(8):545–550.

393. Hayes JP, Lambourn L, Hopkirk JA, Durham SR, Taylor AJ: **Occupational asthma due to styrene.** *Thorax*1991,**46**(5):396–397.

394. Moscato G, Biscaldi G, Cottica D, Pugliese F, Candura S, Candura F: **Occupational asthma due to styrene: two case reports.** *J Occup Med* 1987,**29**(12):957–960.

395. Andersson E, Knutsson A, Hagberg S, Nilsson T, Karlsson B, Alfredsson L, Toren K: **Incidence of asthma among workers exposed to sulphur dioxide and other irritant gases.** *Eur Respir J* 2006,**27**(4):720–725.

396. Koksal N, Hasanoglu HC, Gokirmak M, Yildirim Z, Gultek A: **Apricot sulfurization: an occupation that induces an asthma-like syndrome in agricultural environments.** *Am J Ind Med* 2003,**43**(4):447–453.

397. Harkonen H, Nordman H, Korhonen O, Winblad I: **Long-term effects of exposure to sulfur dioxide. Lung function four years after a pyrite dust explosion.** *Am Rev Respir Dis* 1983,**128**(5):890–893.

398. Rosberg M: **Asthma bronchiale caused by sulfthiazole.** *Acta Medica Scandinavica* 1946, **126:** 185–190.

399. Eriksson KA, Levin JO, Sandstrom T, Lindstrom-Espeling K, Linden G, Stjernberg NL: **Terpene exposure and respiratory effects among workers in Swedish joinery shops.** *Scand J Work Environ Health* 1997,**23**(2):114–120.

400. Seaton A, Cherrie B, Turnbull J: **Rubber glove asthma.** *Br Med J (Clin Res Ed)* 1988,**296**(6621):531–532.

401. Honda I, Kohrogi H, Ando M, Araki S, Ueno T, Futatsuka M, Ueda A: **Occupational asthma induced by the fungicide tetrachloroisophthalonitrile.** *Thorax* 1992,**47**(9):760–761.

402. Baur X, Bittner C: **Occupational obstructive airway diseases caused by the natural gas odorant tetrahydrothiophene–two case reports.** *Am J Ind Med* 2009,**52**(12):982–986.

403. Vandenplas O, Delwiche JP, Auverdin J, Caroyer UM, Cangh FB: **Asthma to tetramethrin.** *Allergy*2000,**55**(4):417–418.

404. Shelton D, Urch B, Tarlo SM: **Occupational asthma induced by a carpet fungicide–tributyl tin oxide.** *J Allergy Clin Immunol* 1992,**90**(2):274–275.

405. Bruckner HC: **Extrinsic asthma in a tungsten carbide worker.** *J Occup Med* 1967,**9**(10):518–519.

406. Lee HS, Wang YT, Yeo CT, Tan KT, Ratnam KV: **Occupational asthma due to tylosin tartrate.** *Br J Ind Med*1989,**46**(7):498–499.

407. Frigas E, Filley WV, Reed CE: **Asthma induced by dust from urea-formaldehyde foam insulating material.***Chest* 1981,**79**(6):706–707.

408. Bertrand JP, Simon V, Chau N: **Associations of symptoms related to isocyanate, ureaformol, and formophenolic exposures with respiratory symptoms and lung function in coal miners.** *Int J Occup Environ Health* 2007,**13**(2):181–187.

409. Irsigler GB, Visser PJ, Spangenberg PA: **Asthma and chemical bronchitis in vanadium plant workers.** *Am J Ind Med* 1999,**35**(4):366–374.

410. Pistelli R, Pupp N, Forastiere F, Agabiti N, Corbo GM, Tidei F, Perucci CA: **Increase of nonspecific bronchial reactivity after occupational exposure to vanadium.** *Med Lav* 1991,**82**(3):270–275.

411. Kielkowski D, Rees D: *Report on exposure and health, assessment of vanadium workers.* 1997.

412. Musk AW, Tees JG: **Asthma caused by occupational exposure to vanadium compounds.** *Med J Aust*1982,**1**(4):183–184.

413. Malo JL, Cartier A: **Occupational asthma due to fumes of galvanized metal.** *Chest* 1987,**92**(2):375–377.

414. Malo JL, Cartier A, Dolovich J: **Occupational asthma due to zinc.** *Eur Respir J* 1993,**6**(3):447–450.

415. Kawane H, Soejima R, Umeki S, Niki Y: **Metal fume fever and asthma.** *Chest* 1988,**93**(5):1116–1117.

416. Hambrook DW, Fink JN: **Airbag asthma: a case report and review of the literature.** *Ann Allergy Asthma Immunol* 2006,**96**(2):369–372.

417. Yang CY, Huang CC, Chiu HF, Chiu JF, Lan SJ, Ko YC: **Effects of occupational dust exposure on the respiratory health of Portland cement workers.** *J Toxicol Environ Health* 1996,**49**(6):581–588.

418. Abrons HL, Petersen MR, Sanderson WT, Engelberg AL, Harber P: **Symptoms, ventilatory function, and environmental exposures in Portland cement workers.** *Br J Ind Med* 1988,**45**(6):368–375.

419. Fell AK, Thomassen TR, Kristensen P, Egeland T, Kongerud J: **Respiratory symptoms and ventilatory function in workers exposed to portland cement dust.** *J Occup Environ Med* 2003,**45**(9):1008–1014.

420. Al-Neaimi YI, Gomes J, Lloyd OL: **Respiratory illnesses and ventilatory function among workers at a cement factory in a rapidly developing country.** *Occup Med (Lond)* 2001,**51**(6):367–373.

421. Ali BA, Ballal SG, Albar AA, Ahmed HO: **Post-shift changes in pulmonary function in a cement factory in eastern Saudi Arabia.** *Occup Med (Lond)* 1998,**48**(8):519–522.

422. Kalacic I: **Chronic nonspecific lung disease in cement workers.** *Arch Environ Health* 1973,**26**(2):78–83.

423. Mengesha YA, Bekele A: **Relative chronic effects of different occupational dusts on respiratory indices and health of workers in three Ethiopian factories.** *Am J Ind Med* 1998,**34**(4):373–380.

424. Lockman LE: **Case report: allergic contact dermatitis and new-onset asthma. Chromium exposure during leather tanning.** *Can Fam Physician* 2002, **48:** 1907–1909.

425. De Raeve H, Vandecasteele C, Demedts M, Nemery B: **Dermal and respiratory sensitization to chromate in a cement floorer.** *Am J Ind Med* 1998,**34**(2):169–176.

426. Shirakawa T, Morimoto K: **Brief reversible bronchospasm resulting from bichromate exposure.** *Arch Environ Health* 1996,**51**(3):221–226.

427. Nordby KC, Fell AK, Noto H, Eduard W, Skogstad M, Thomassen Y, Bergamaschi A, Kongerud J, Kjuus H:**Exposure to thoracic**

dust, airway symptoms and lung function in cement production workers. *Eur Respir J* 2011,**38**(6):1278–1286.

428. Piirila P, Espo T, Pfaffli P, Riihimaki V, Wolff H, Nordman H: **Prolonged respiratory symptoms caused by thermal degradation products of freons.** *Scand J Work Environ Health* 2003,**29**(1):71–77.

429. Lee J, Lee C, Kim CH: **Uncontrolled occupational exposure to 1,1-dichloro-1-Fluoroethane (HCFC-141b) is associated with acute pulmonary toxicity.** *Chest* 2009,**135**(1):149–155.

430. Mirabelli MC, Zock JP, Plana E, Anto JM, Benke G, Blanc PD, Dahlman-Hoglund A, Jarvis DL, Kromhout H, Lillienberg L, *et al.*: **Occupational risk factors for asthma among nurses and related healthcare professionals in an international study.** *Occup Environ Med* 2007,**64**(7):474–479.

431. Rosenman KD, Reilly MJ, Schill DP, Valiante D, Flattery J, Harrison R, Reinisch F, Pechter E, Davis L, Tumpowsky CM, *et al.*: **Cleaning products and work-related asthma.** *J Occup Environ Med* 2003,**45**(5):556–563.

432. Tabar AI, Alvarez MJ, Acero S, Olaguibel JM, Garcia BE, Quirce S: **Reactive airways dysfunction syndrome: two case reports.** *J Investig Allergol Clin Immunol* 1998,**8**(2):119–122.

433. Vizcaya D, Mirabelli MC, Anto JM, Orriols R, Burgos F, Arjona L, Zock JP: **A workforce-based study of occupational exposures and asthma symptoms in cleaning workers.** *Occup Environ Med* 2011,**68**(12):914–919.

434. Murphy DM, Fairman RP, Lapp NL, Morgan WK: **Severe airway disease due to inhalation of fumes from cleansing agents.** *Chest* 1976,**69**(3):372–376.

435. Mapp CE, Beghe B, Balboni A, Zamorani G, Padoan M, Jovine L, Baricordi OR, Fabbri LM: **Association between HLA genes and susceptibility to toluene diisocyanate-induced asthma.** *Clin Exp Allergy*2000,**30**(5):651–656.

436. Jones RN, Hughes JM, Lehrer SB, Butcher BT, Glindmeyer HW, Diem JE, Hammad YY, Salvaggio J, Weill H: **Lung function consequences of exposure and hypersensitivity in workers**

who process green coffee beans. *Am Rev Respir Dis* 1982,**125**(2):199–202.

437. Zuskin E, Kanceljak B, Skuric Z, Butkovic D: **Bronchial reactivity in green coffee exposure.** *Br J Ind Med*1985,**42**(6):415–420.

438. Lemiere C, Malo JL, McCants M, Lehrer S: **Occupational asthma caused by roasted coffee: immunologic evidence that roasted coffee contains the same antigens as green coffee, but at a lower concentration.** *J Allergy Clin Immunol* 1996,**98**(2):464–466.

439. Johansen JP, Viskum S: **Asthma associated with the handling of green coffee beans.** *Ugeskr Laeger*1987,**149**(42):2853.

440. Oldenburg M, Bittner C, Baur X: **Health risks due to coffee dust.** *Chest* 2009,**136**(2):536–544.

441. Christiani DC, Wang XR, Pan LD, Zhang HX, Sun BX, Dai H, Eisen EA, Wegman DH, Olenchock SA: **Longitudinal changes in pulmonary function and respiratory symptoms in cotton textile workers. A 15-yr follow-up study.** *Am J Respir Crit Care Med* 2001,**163**(4):847–853.

442. Latza U, Oldenburg M, Baur X: **Endotoxin exposure and respiratory symptoms in the cotton textile industry.** *Arch Environ Health* 2004,**59**(10):519–525.

443. Oldenburg M, Latza U, Baur X: **Exposure-response relationship between endotoxin exposure and lung function impairment in cotton textile workers.** *Int Arch Occup Environ Health* 2007,**80**(5):388–395.

444. Hayes GB, Ye TT, Lu PL, Dai HL, Christiani DC: **Respiratory disease in cotton textile workers: epidemiologic assessment of small airway function.** *Environ Res* 1994,**66**(1):31–43.

445. Wang XR, Pan LD, Zhang HX, Sun BX, Dai HL, Christiani DC: **A longitudinal observation of early pulmonary responses to cotton dust.** *Occup Environ Med* 2003,**60**(2):115–121.

446. Li D, Zhong YN, Rylander R, Ma QY, Zhou XY: **Longitudinal study of the health of cotton workers.** *Occup Environ Med* 1995,**52**(5):328–331.

447. Mberikunashe J, Banda S, Chadambuka A, Gombe NT, Shambira G, Tshimanga M, Matchaba-Hove R:**Prevalence and risk factors for obstructive respiratory conditions among textile industry workers in Zimbabwe, 2006.** *Pan Afr Med J* 2010, **6:** 1.

448. Hendy MS, Beattie BE, Burge PS: **Occupational asthma due to an emulsified oil mist.** *Br J Ind Med*1985,**42**(1):51–54.

449. Keskinen H, Nordman H, Terho EO: **ECG ink as a cause of asthma.** *Allergy* 1981,**36**(4):275–276.

450. Rodenstein D, Stanescu DC: **Bronchial asthma following exposure to ECG ink.** *Ann Allergy* 1982,**48**(6):351–352.

451. Milton DK, Wypij D, Kriebel D, Walters MD, Hammond SK, Evans JS: **Endotoxin exposure-response in a fiberglass manufacturing facility.** *Am J Ind Med* 1996,**29**(1):3–13.

452. Jaakkola JJ, Piipari R, Jaakkola MS: **Occupation and asthma: a population-based incident case–control study.***Am J Epidemiol* 2003,**158**(10):981–987.

453. Lambourn EM, Hayes JP, McAllister WA, Taylor AJ: **Occupational asthma due to EPO 60.** *Br J Ind Med*1992,**49**(4):294–295.

454. Cockcroft DW, Cartier A, Jones G, Tarlo SM, Dolovich J, Hargreave FE: **Asthma caused by occupational exposure to a furan-based binder system.** *J Allergy Clin Immunol* 1980,**66**(6):458–463.

455. Chan-Yeung M, Schulzer M, MacLean L, Dorken E, Grzybowski S: **Epidemiologic health survey of grain elevator workers in British Columbia.** *Am Rev Respir Dis* 1980,**121**(2):329–338.

456. Williams N, Skoulas A, Merriman JE: **Exposure to Grain Dust. I. a Survey of the Effects.** *J Occup Med* 1964, **6:**319–329.

457. Skoulas A, Williams N, Merriman JE: **Exposure to Grain Dust. Ii. a Clinical Study of the Effects.** *J Occup Med*1964, **6:** 359–372.

458. Baur X, Preisser A, Wegner R: **Asthma due to grain dust.** *Pneumologie* 2003,**57**(6):335–339.

459. Chan-Yeung M, Dimich-Ward H, Enarson DA, Kennedy SM: **Five cross-sectional studies of grain elevator workers.** *Am J Epidemiol* 1992,**136**(10):1269–1279.

460. Ye TT, Huang JX, Shen YE, Lu PL, Christiani DC: **Respiratory symptoms and pulmonary function among Chinese rice-granary workers.** *Int J Occup Environ Health* 1998,**4**(3):155–159.

461. Dube D, Puruckherr M, Byrd RP Jr, Roy TM: **Reactive airways dysfunction syndrome following metal fume fever.** *Tenn Med* 2002,**95**(6):236–238.

462. Zacharisen MC, Kadambi AR, Schlueter DP, Kurup VP, Shack JB, Fox JL, Anderson HA, Fink JN: **The spectrum of respiratory disease associated with exposure to metal working fluids.** *J Occup Environ Med*1998,**40**(7):640–647.

463. Wieslander G, Janson C, Norback D, Bjornsson E, Stalenheim G, Edling C: **Occupational exposure to water-based paints and self-reported asthma, lower airway symptoms, bronchial hyperresponsiveness, and lung function.** *Int Arch Occup Environ Health* 1994,**66**(4):261–267.

464. Temel O, Sakar Coskun A, Yaman N, Sarioglu N, Alkac C, Konyar I, Ozgen Alpaydin A, Celik P, Cengiz Ozyurt B, Keskin E, *et al.*: **Occupational asthma in welders and painters.** *Tuberk Toraks* 2010,**58**(1):64–70.

465. Baur X, Schneider EM, Wieners D, Czuppon AB: **Occupational asthma to perfume.** *Allergy* 1999,**54**(12):1334–1335.

466. Lings S: **Pesticide lung: a pilot investigation of fruit-growers and farmers during the spraying season.** *Br J Ind Med* 1982,**39**(4):370–376.

467. Boers D, van Amelsvoort L, Colosio C, Corsini E, Fustinoni S, Campo L, Bosetti C, La Vecchia C, Vergieva T, Tarkowski M, *et al.*: **Asthmatic symptoms after exposure to ethylenebisdithiocarbamates and other pesticides in the Europit field studies.** *Hum Exp Toxicol* 2008,**27**(9):721–727.

468. Ng TP, Lee HS, Malik MA, Chee CB, Cheong TH, Wang YT: **Asthma in chemical workers exposed to aliphatic polyamines.** *Occup Med (Lond)* 1995,**45**(1):45–48.

469. Cartier A, Vandenplas O, Grammer LC, Shaughnessy MA, Malo JL: **Respiratory and systemic reaction following exposure to heated electrostatic polyester paint.** *Eur Respir J* 1994,**7**(3):608–611.

470. Zuskin E, Mustajbegovic J, Schachter EN, Kern J, Budak A, Godnic-Cvar J: **Respiratory findings in synthetic textile workers.** *Am J Ind Med* 1998,**33**(3):263–273.

471. Burge PS, Scott JA, McCoach J: **Occupational asthma caused by aluminum.** *Allergy* 2000,**55**(8):779–780.

472. Chan-Yeung M, Wong R, MacLean L, Tan F, Schulzer M, Enarson D, Martin A, Dennis R, Grzybowski S:**Epidemiologic health study of workers in an aluminum smelter in British Columbia. Effects on the respiratory system.** *Am Rev Respir Dis* 1983,**127**(4):465–469.

473. Desjardins A, Bergeron JP, Ghezzo H, Cartier A, Malo JL: **Aluminium potroom asthma confirmed by monitoring of forced expiratory volume in one second.** *Am J Respir Crit Care Med* 1994,**150**(6 Pt 1):1714–1717.

474. Musk AW, de Klerk NH, Beach JR, Fritschi L, Sim MR, Benke G, Abramson M, McNeil JJ: **Respiratory symptoms and lung function in alumina refinery employees.** *Occup Environ Med* 2000,**57**(4):279–283.

475. O'Donnell TV, Welford B, Coleman ED: **Potroom asthma: New Zealand experience and follow-up.** *Am J Ind Med* 1989,**15**(1):43–49.

476. Saric M, Godnic-Cvar J, Gomzi M, Stilinovic L: **The role of atopy in potroom workers' asthma.** *Am J Ind Med*1986,**9**(3):239–242.

477. Sorgdrager B, de Looff AJ, de Monchy JG, Pal TM, Dubois AE, Rijcken B: **Occurrence of occupational asthma in aluminum potroom workers in relation to preventive measures.** *Int Arch Occup Environ Health* 1998, **71:**53–59.

478. Sorgdrager B, de Looff AJ, Pal TM, van Dijk FJ, de Monchy JG: **Factors affecting FEV1 in workers with potroom asthma after their removal from exposure.** *Int Arch Occup Environ Health* 2001,**74**(1):55–58.

479. Blomqvist A, Duzakin-Nystedt M, Ohlson CG, Andersson L, Jonsson B, Nielsen J, Welinder H: **Airways symptoms, immunological response and exposure in powder painting.** *Int Arch Occup Environ Health*2005,**78**(2):123–131.

480. Nakano Y, Tsuchiya T, Hirose K, Chida K: **Occupational asthma caused by pyrazolone derivative used in silver halide photographic paper.** *Chest* 2000,**118**(1):246–248.

481. Alanko K, Keskinen H, Bjorksten F, Ojanen S: **Immediate-type hypersensitivity to reactive dyes.** *Clin Allergy*1978,**8**(1):25–31.

482. Nilsson R, Nordlinder R, Wass U, Meding B, Belin L: **Asthma, rhinitis, and dermatitis in workers exposed to reactive dyes.** *Br J Ind Med* 1993,**50**(1):65–70.

483. Park HS, Lee MK, Hong CS: **Reactive dye induced occupational asthma without nonspecific bronchial hyperreactivity.** *Yonsei Med J* 1990,**31**(2):98–102.

484. Park HS, Lee MK, Kim BO, Lee KJ, Roh JH, Moon YH, Hong CS: **Clinical and immunologic evaluations of reactive dye-exposed workers.** *J Allergy Clin Immunol* 1991,**87**(3):639–649.

485. Romano C, Sulotto F, Pavan I, Chiesa A, Scansetti G: **A new case of occupational asthma from reactive dyes with severe anaphylactic response to the specific challenge.** *Am J Ind Med* 1992,**21**(2):209–216.

486. Forrester BG: **Reactive airways dysfunction syndrome: occurrence after exposure to a refractory ceramic fiber-phosphoric acid binder mixture.** *South Med J* 1997,**90**(4):447–450.

487. Lemasters GK, Lockey JE, Levin LS, McKay RT, Rice CH, Horvath EP, Papes DM, Lu JW, Feldman DJ: **An industry-wide pulmonary study of men and women manufacturing refractory ceramic fibers.** *Am J Epidemiol* 1998,**148**(9):910–919.

488. Almeida AG, Duarte R, Mieiro L, Paiva AC, Rodrigues AM, Almeida MH, Barbara C: **Pulmonary function in Portuguese firefighters.** *Rev Port Pneumol* 2007,**13**(3):349–364.

489. Moisan TC: **Prolonged asthma after smoke inhalation: a report of three cases and a review of previous reports.** *J Occup Med* 1991,**33**(4):458–461.

490. Greven F, Krop E, Spithoven J, Rooyackers J, Kerstjens H, Heederik D: **Lung function, bronchial hyperresponsiveness, and atopy among firefighters.** *Scand J Work Environ Health* 2011,**37**(4):325–331.

491. Kelsall HL, Sim MR, Forbes AB, Glass DC, McKenzie DP, Ikin JF, Abramson MJ, Blizzard L, Ittak P: **Symptoms and medical conditions in Australian veterans of the 1991 Gulf War: relation to immunisations and other Gulf War exposures.** *Occup Environ Med* 2004,**61**(12):1006–1013.

492. Ekici A, Ekici M, Kurtipek E, Akin A, Arslan M, Kara T, Apaydin Z, Demir S: **Obstructive airway diseases in women exposed to biomass smoke.** *Environ Res* 2005,**99**(1):93–98.

493. Burge PS, Perks W, O'Brien IM, Hawkins R, Green M: **Occupational asthma in an electronics factory.** *Thorax*1979,**34**(1):13–18.

494. Lee HS, Koh D, Chia HP, Phoon WH: **Symptoms, lung function, and diurnal variation in peak expiratory flow rate among female solderers in the electronics industry.** *Am J Ind Med* 1994,**26**(5):613–619.

495. Stevens JJ: **Asthma due to soldering flux: a polyether alcohol-polypropylene glycol mixture.** *Ann Allergy*1976,**36**(6):419–422.

496. Cakmak A, Ekici A, Ekici M, Arslan M, Iteginli A, Kurtipek E, Kara T: **Respiratory findings in gun factory workers exposed to solvents.** *Respir Med* 2004,**98**(1):52–56.

497. Ebbehoj NE, Hein HO, Suadicani P, Gyntelberg F: **Occupational organic solvent exposure, smoking, and prevalence of chronic bronchitis-an epidemiological study of 3387 men.** *J Occup Environ Med*2008,**50**(7):730–735.

498. Tarlo SM: **Occupational asthma induced by tall oil in the rubber tyre industry.** *Clin Exp Allergy*1992,**22**(1):99–101.

499. Bayeux-Dunglas MC, Deparis P, Touati MA, Ameille J: **Occupational asthma in a teacher after repeated exposure to tear gas.** *Rev Mal Respir* 1999,**16**(4):558–559.

500. Hill AR, Silverberg NB, Mayorga D, Baldwin HE: **Medical hazards of the tear gas CS. A case of persistent, multisystem, hypersensitivity reaction and review of the literature.** *Medicine (Baltimore)* 2000,**79**(4):234–240.

501. Hu H, Christiani D: **Reactive airways dysfunction after exposure to teargas.** *Lancet* 1992,**339**(8808):1535.

502. Roth VS, Franzblau A: **RADS after exposure to a riot-control agent: a case report.** *J Occup Environ Med*1996,**38**(9):863–865.

503. El-Zein M, Malo JL, Infante-Rivard C, Gautrin D: **Incidence of probable occupational asthma and changes in airway calibre and responsiveness in apprentice welders.** *Eur Respir J* 2003,**22**(3):513–518.

504. Jafari AJ, Assari MJ: **Respiratory effects from work-related exposure to welding fumes in Hamadan, Iran.***Arch Environ Health* 2004,**59**(3):116–120.

505. Nakadate T, Aizawa Y, Yagami T, Zheg YQ, Kotani M, Ishiwata K: **Change in obstructive pulmonary function as a result of cumulative exposure to welding fumes as determined by magnetopneumography in Japanese arc welders.** *Occup Environ Med* 1998,**55**(10):673–677.

506. Beach JR, Dennis JH, Avery AJ, Bromly CL, Ward RJ, Walters EH, Stenton SC, Hendrick DJ: **An epidemiologic investigation of asthma in welders.** *Am J Respir Crit Care Med* 1996,**154**(5):1394–1400.

507. Hannu T, Piipari R, Toskala E: **Immediate hypersensitivity type of occupational laryngitis in a welder exposed to welding fumes of stainless steel.** *Am J Ind Med* 2006,**49**(5):402–405.

508. Karjalainen A, Martikainen R, Oksa P, Saarinen K, Uitti J: **Incidence of asthma among Finnish construction workers.** *J Occup Environ Med* 2002,**44**(8):752–757.

509. Contreras GR, Chan-Yeung M: **Bronchial reactions to exposure to welding fumes.** *Occup Environ Med*1997,**54**(11):836–839.

510. Hannu T, Piipari R, Kasurinen H, Keskinen H, Tuppurainen M, Tuomi T: **Occupational asthma due to manual metal-arc welding of special stainless steels.** *Eur Respir J* 2005,**26**(4):736–739.

511. Vandenplas O, Delwiche JP, Vanbilsen ML, Joly J, Roosels D: **Occupational asthma caused by aluminium welding.** *Eur Respir J* 1998,**11**(5):1182–1184.

512. Vandenplas O, Dargent F, Auverdin JJ, Boulanger J, Bossiroy JM, Roosels D, Van de Weyer R: **Occupational asthma due to gas**

metal arc welding on mild steel. *Thorax* 1995,**50**(5):587–588. discussion 589

513. Keskinen H, Kalliomaki PL, Alanko K: **Occupational asthma due to stainless steel welding fumes.** *Clin Allergy*1980,**10**(2):151–159.

514. Halvani GH, Zare M, Halvani A, Barkhordari A: **Evaluation and comparison of respiratory symptoms and lung capacities in tile and ceramic factory workers of Yazd.** *Arh Hig Rada Toksikol* 2008,**59**(3):197–204.

515. Neghab M, Zadeh JH, Fakoorziba MR: **Respiratory toxicity of raw materials used in ceramic production.** *Ind Health* 2009,**47**(1):64–69.

516. Oliver LC, Miracle-McMahill H, Littman AB, Oakes JM, Gaita RR Jr: **Respiratory symptoms and lung function in workers in heavy and highway construction: a cross-sectional study.** *Am J Ind Med* 2001,**40**(1):73–86.

517. Kaukiainen A, Martikainen R, Riala R, Reijula K, Tammilehto L: **Work tasks, chemical exposure and respiratory health in construction painting.** *Am J Ind Med* 2008,**51**(1):1–8.

518. Walusiak J, Krawczyk-Adamus P, Hanke W, Wittczak T, Palczynski C: **Small nonspecialized farming as a protective factor against immediate-type occupational respiratory allergy?** *Allergy* 2004,**59**(12):1294–1300.

519. Gomez MI, Hwang SA, Lin S, Stark AD, May JJ, Hallman EM: **Prevalence and predictors of respiratory symptoms among New York farmers and farm residents.** *Am J Ind Med* 2004,**46**(1):42–54.

520. Dosman JA, Graham BL, Hall D, Van Loon P, Bhasin P, Froh F: **Respiratory symptoms and pulmonary function in farmers.** *J Occup Med* 1987,**29**(1):38–43.

521. Rask-Andersen A: **Asthma increase among farmers: a 12-year follow-up.** *Ups J Med Sci* 2011,**116**(1):60–71.

522. Portengen L, Preller L, Tielen M, Doekes G, Heederik D: **Endotoxin exposure and atopic sensitization in adult pig farmers.** *J Allergy Clin Immunol* 2005,**115**(4):797–802.

523. Monso E, Riu E, Radon K, Magarolas R, Danuser B, Iversen M, Morera J, Nowak D: **Chronic obstructive pulmonary disease in never-smoking animal farmers working inside confinement buildings.** *Am J Ind Med*2004,**46**(4):357–362.

524. Radon K, Weber C, Iversen M, Danuser B, Pedersen S, Nowak D: **Exposure assessment and lung function in pig and poultry farmers.** *Occup Environ Med* 2001,**58**(6):405–410.

525. Hansen ES: **A cohort mortality study of foundry workers.** *Am J Ind Med* 1997,**32**(3):223–233.

526. Hahn R, Beck B: **Incidence of chronic bronchitis in foundry workers.** *Z Erkr Atmungsorgane*1986,**166**(3):267–273.

527. Pechter E, Davis LK, Tumpowsky C, Flattery J, Harrison R, Reinisch F, Reilly MJ, Rosenman KD, Schill DP, Valiante D, *et al.*: **Work-related asthma among health care workers: surveillance data from California, Massachusetts, Michigan, and New Jersey, 1993–1997.** *Am J Ind Med* 2005,**47**(3):265–275.

528. Arif AA, Delclos GL: **Association between cleaning-related chemicals and work-related asthma and asthma symptoms among healthcare professionals.** *Occup Environ Med* 2012,**69**(1):35–40.

529. Arif AA, Delclos GL, Serra C: **Occupational exposures and asthma among nursing professionals.** *Occup Environ Med* 2009,**66**(4):274–278.

530. Bala S, Tabaku A: **Chronic obstructive pulmonary disease in iron-steel and ferrochrome industry workers.***Cent Eur J Public Health* 2010,**18**(2):93–98.

531. Rylander R, Carvalheiro MF: **Airways inflammation among workers in poultry houses.** *Int Arch Occup Environ Health* 2006,**79**(6):487–490.

532. Hagmar L, Schutz A, Hallberg T, Sjoholm A: **Health effects of exposure to endotoxins and organic dust in poultry slaughter-house workers.** *Int Arch Occup Environ Health* 1990,**62**(2):159–164.

533. Danuser B, Wyss C, Hauser R, von Planta U, Folsch D: **Lung function and symptoms in employees of poultry farms.** *Soz Praventivmed* 1988,**33**(6):286–291.

534. King BS, Page EH, Mueller CA, Dollberg DD, Gomez KE, Warren AM: **Eye and respiratory symptoms in poultry processing workers exposed to chlorine by-products.** *Am J Ind Med* 2006,**49**(2):119–126.

535. Borghetti C, Magarolas R, Badorrey I, Radon K, Morera J, Monso E: **Sensitization and occupational asthma in poultry workers.** *Med Clin (Barc)* 2002,**118**(7):251–255.

536. Perfetti L, Cartier A, Malo JL: **Occupational asthma in poultry-slaughterhouse workers.** *Allergy*1997,**52**(5):594–595.

537. Vogelzang PF, van der Gulden JW, Folgering H, Heederik D, Tielen MJ, van Schayck CP: **Longitudinal changes in bronchial responsiveness associated with swine confinement dust exposure.** *Chest* 2000,**117**(5):1488–1495.

538. Reynolds SJ, Donham KJ, Whitten P, Merchant JA, Burmeister LF, Popendorf WJ: **Longitudinal evaluation of dose–response relationships for environmental exposures and pulmonary function in swine production workers.** *Am J Ind Med* 1996,**29**(1):33–40.

539. Schwartz DA, Donham KJ, Olenchock SA, Popendorf WJ, Van Fossen DS, Burmeister LF, Merchant JA:**Determinants of longitudinal changes in spirometric function among swine confinement operators and farmers.** *Am J Respir Crit Care Med* 1995,**151**(1):47–53.

540. Vogelzang PF, van der Gulden JW, Folgering H, Kolk JJ, Heederik D, Preller L, Tielen MJ, van Schayck CP:**Endotoxin exposure as a major determinant of lung function decline in pig farmers.** *Am J Respir Crit Care Med* 1998,**157**(1):15–18.

541. Banauch GI, Alleyne D, Sanchez R, Olender K, Cohen HW, Weiden M, Kelly KJ, Prezant DJ: **Persistent hyperreactivity and reactive airway dysfunction in firefighters at the World Trade Center.** *Am J Respir Crit Care Med* 2003,**168**(1):54–62.

542. Banauch GI, Dhala A, Prezant DJ: **Pulmonary disease in rescue workers at the World Trade Center site.** *Curr Opin Pulm Med* 2005,**11**(2):160–168.

543. Banauch GI, Hall C, Weiden M, Cohen HW, Aldrich TK, Christodoulou V, Arcentales N, Kelly KJ, Prezant DJ:**Pulmonary function after exposure to the World Trade Center collapse in the New York City Fire Department.** *Am J Respir Crit Care Med* 2006,**174**(3):312–319.

544. Herbstman JB, Frank R, Schwab M, Williams DL, Samet JM, Breysse PN, Geyh AS: **Respiratory effects of inhalation exposure among workers during the clean-up effort at the World Trade Center disaster site.***Environ Res* 2005,**99**(1):85–92.

545. Salzman SH, Moosavy FM, Miskoff JA, Friedmann P, Fried G, Rosen MJ: **Early respiratory abnormalities in emergency services police officers at the World Trade Center site.** *J Occup Environ Med* 2004,**46**(2):113–122.

546. Mauer MP, Cummings KR, Hoen R: **Long-term respiratory symptoms in World Trade Center responders.***Occup Med (Lond)* 2010,**60**(2):145–151.

547. Feinstein AR, Horwitz RI: **Problems in the "evidence" of "evidence-based medicine".** *Am J Med*1997,**103**(6):529–535.

548. Miettinen OS: **Evidence in medicine: invited commentary.** *Cmaj* 1998,**158**(2):215–221.

549. Miettinen OS: **Evidence-based medicine, case-based medicine; scientific medicine, quasi-scientific medicine. Commentary on Tonelli (2006), Integrating evidence into clinical practice: an alternative to evidence-based approaches. Journal of Evaluation in Clinical Practice 12, 248–256.** *J Eval Clin Pract*2006,**12**(3):260–264.

550. Tobin MJ: **Counterpoint: evidence-based medicine lacks a sound scientific base.** *Chest* 2008,**133**(5):1071–1074. discussion 1074–1077

551. Lijmer JG, Mol BW, Heisterkamp S, Bonsel GJ, Prins MH, van der Meulen JH, Bossuyt PM: **Empirical evidence of design-related bias in studies of diagnostic tests.** *Jama* 1999,**282**(11):1061–1066.

552. Beach J, Rowe BH, Blitz S, Crumley E, Hooton N, Russell K, Spooner C, Klassen T: *Diagnosis and management of work-related asthma. Evidence Report/Technology Assessment No 129*. U.S. Department of Health and Human Services, Agency for Healthcare Research and Quality, Rockville; 2005. http://www.ahrq.gov/downloads/pub/evidence/pdf/asthmawork/asthwork.pdf

553. Kunz R, Burnand B, Schunemann HJ: **The GRADE System. An international approach to standardize the graduation of evidence and recommendations in guidelines.** *Internist (Berl)* 2008,**49**(6):673–680.

554. West S, King V, CT S, Lohr KN, McKoy N, Sutton SF, Lux L: *Systems to rate the strength of scientific evidence, vol. 47*. Agency for Healthcare, Research and Quality, Rockville; 2002.

555. Quint J, Beckett WS, Campleman SL, Sutton P, Prudhomme J, Flattery J, Harrison R, Cowan B, Kreutzer R:**Primary prevention of occupational asthma: identifying and controlling exposures to asthma-causing agents.** *Am J Ind Med* 2008,**51**(7):477–491.

556. AbuDhaise BA, Rabi AZ, Al Zwairy MA, El Hader AF, El Qaderi S: **Pulmonary manifestations in cement workers in Jordan.** *Int J Occup Med Environ Health* 1997,**10**(4):417–428.

557. Salvi SS, Barnes PJ: **Chronic obstructive pulmonary disease in non-smokers.** *Lancet* 2009,**374**(9691):733–743.

558. Blanc FX, Coirault C, Oliviero P, Lecarpentier Y: **Relaxation of tracheal smooth muscle is impaired in innate airway hyperresponsiveness.** *Eur Respir J* 2009,**34**(2):417–424.

559. Balmes JR, Nowak D: **COPD caused by occupational exposure.** In *COPD*. Edited by: Donner CF, Carone M. Clinical publishing, Oxford; 2007:85–95.

560. Blanc PD, Toren K: **Occupation in chronic obstructive pulmonary disease and chronic bronchitis: an update.** *Int J Tuberc Lung Dis* 2007,**11**(3):251–257..

561. Toren K, Balmes J: **Chronic obstructive pulmonary disease: does occupation matter?** *Am J Respir Crit Care Med* 2007,**176**(10):951–952.

562. Bardana EJ Jr: **Reactive airways dysfunction syndrome (RADS): guidelines for diagnosis and treatment and insight into likely prognosis.** *Ann Allergy Asthma Immunol* 1999,**83**(6 Pt 2):583–586.

563. Vandenplas O, Malo JL: **Definitions and types of work-related asthma: a nosological approach.** *Eur Respir J*2003,**21**(4):706–712.

564. Blanc PD, Menezes AM, Plana E, Mannino DM, Hallal PC, Toren K, Eisner MD, Zock JP: **Occupational exposures and COPD: an ecological analysis of international data.** *Eur Respir J* 2009,**33**(2):298–304.

565. Malo JL, Chan-Yeung M: **Agents causing occupational asthma.** *J Allergy Clin Immunol* 2009,**123**(3):545–550.

566. Cullinan P: **Irritant-induced asthma from work. What happens next?** *Am J Respir Crit Care Med*2009,**179**(10):857–858.

567. Tarlo SM, Boulet LP, Cartier A, Cockcroft D, Cote J, Hargreave FE, Holness L, Liss G, Malo JL, Chan-Yeung M:**Canadian Thoracic Society guidelines for occupational asthma.** *Can Respir J* 1998,**5**(4):289–300.

568. Ernst P, Fitzgerald JM, Spier S: **Canadian Asthma Consensus Conference Summary of recommendations.***Can Respir J* 1996,**3**(2):89–101.

569. Newman Taylor AJ, Cullinan P, Burge PS, Nicholson P, Boyle C: **BOHRF guidelines for occupational asthma.***Thorax* 2005,**60**(5):364–366.

570. Heederik D, van Rooy F: **Exposure assessment should be integrated in studies on the prevention and management of occupational asthma.** *Occup Environ Med* 2008,**65**(3):149–150.

571. Venables KM: **Prevention of occupational asthma.** *Eur Respir J* 1994,**7**(4):768–778.

572. Vineis P: **Evidence-based primary prevention?** *Scand J Work Environ Health* 2000,**26**(5):443–448.

573. Drummond MF, Stoddart GL, Torrance GW: *Method of economic evaluation of health care programes*. Oxford Medical Publications, Oxford; 1987.

574. Heederik D, Henneberger PK, Redlich CA: **Primary prevention: exposure reduction, skin exposure and respiratory protection.** *Eur Respir Rev* 2012,**21**(124):112–124.

575. Schoenbach V, Wayne D: *Understanding the fundamentals of epidemiology: an evolving text. Chapter 8: Analytic study designs*. epidemiolog

Chapter 7

Occupational Exposure to Asbestos and Lung Cancer in Men: Evidence From a Population-Based Case-Control Study in Eight Canadian Provinces

Paul J Villeneuve[1,2,3*], Marie-Élise Parent[4], Shelley A Harris[2,3,5,6], Kenneth C Johnson[7] and The Canadian Cancer Registries Epidemiology Research Group

[1]*Population Studies Division, Health Canada, Ottawa, Ontario, Canada*
[2]*Division of Occupational and Environmental Health, Dalla Lana School of Public Health, University of Toronto, Toronto, Canada*
[3]*Occupational Cancer Research Centre, Toronto, Ontario, Canada.*
[4]*INRS-Institut Armand-Frappier, University of Quebec, Laval, Quebec, Canada.*
[5]*Division of Epidemiology, Dalla Lana School of Public Health, University of Toronto, Toronto, Canada.*
[6]*Prevention and Cancer Control, Cancer Care Ontario, Toronto, Ontario, Canada.*
[7]*Health Promotion and Chronic Disease Prevention Branch, Public Health Agency of Canada, Ottawa, Canada.*

ABSTRACT

Background: Asbestos is classified as a human carcinogen, and studies have consistently demonstrated that workplace exposure to it increases the risk of developing lung cancer. Few studies have evaluated risks in population-based settings where there is a greater variety in the types of occupations, and exposures.

Methods: This was a population based case–control study with 1,681 incident cases of lung cancer, and 2,053 controls recruited from 8 Canadian provinces between 1994 and 1997. Self-reported questionnaires were used to elicit a lifetime occupational history, including general tasks, and information for other risk factors. Occupational hygienists, who were blinded to case–control status, assigned asbestos exposures to each job on the basis of (i) concentration (low, medium, high), (ii) frequency (<5%, 5-30%, and >30% of the time in a normal work week), and (iii) reliability (possible, probable, definite). Logistic regression was used to estimate odds ratios (ORs) and their corresponding 95% confidence intervals (CI).

Results: Those occupationally exposed to (i) low, and (ii) medium or high concentrations of asbestos had ORs for lung cancer of 1.17 (95% CI=0.92 – 1.50) and 2.16 (95% CI=1.21-3.88), respectively, relative to those who were unexposed. Medium or high exposure to asbestos roughly doubled the risk for lung cancer across all three smoking pack-year categories. The joint relationship between smoking and asbestos was consistent with a multiplicative risk model.

Conclusions: Our findings provide further evidence that exposure to asbestos has contributed to an increased risk of lung cancer in Canadian workplaces, and suggests that nearly 3% of lung cancers among Canadian men are caused by occupational exposure to asbestos.

Keywords: Lung cancer; Asbestos; Cigarette smoking; Case–control; Occupational epidemiology

BACKGROUND

Lung cancer continues to be the leading cause of cancer among Canadian men, and in 2012, it was estimated that 13,300 men would be diagnosed with lung cancer and 10,800 would die of it [1]. While cigarette smoking is recognized as the leading cause of lung cancer, many occupational exposures, including asbestos, have also been shown to increase risk. Asbestos is a term used to describe six naturally fibrous minerals, and one of these, chrysotile, accounts for 95% of the asbestos ever used worldwide, and until recently was the only type produced in Canada [2]. All forms of asbestos have long been recognized as human carcinogens by the United States Environmental

Protection Agency [3], the International Agency for Research on Cancer [4], and the National Toxicology Program [5]. This conclusion is based largely on unequivocal evidence assembled from epidemiological studies that have found excesses of lung cancer and mesothelioma in highly exposed textile workers, miners, and cement factory workers [4, 6].

Today, more than 90% of the asbestos produced worldwide is used to manufacture asbestos sheets and pipes [7]. The World Health Organization has estimated that approximately 125 million individuals continue to be exposed to asbestos in the workplace [8]. Occupational exposure to asbestos in Canada has decreased dramatically over the past two decades due to provincial occupational health and safety controls that have been implemented. While those involved in the mining of asbestos are at higher risk of developing asbestos-related disease, the precautions offered to these workers to limit exposure are greater than those unwittingly exposed through other trades. Overall, the mining of asbestos in Canada has decreased dramatically, and in 2011, for the first time in over 130 years, production was halted [9], Today, in Canada, the most common sources of asbestos exposure arise from the repair, renovation, and demolition of older (pre-1980) buildings.

Relatively few studies have examined associations between workplace exposure to asbestos and lung cancer using a population-based design. Population-based designs provide important features that include an ability to estimate risks over a wider range of exposure levels than those typically reported in industry-specific studies. They provide the opportunity to characterize the frequency and nature of exposures in the general population. Moreover, because such studies cover diverse occupational groups, there is a reduced impact of confounders that may be specific to particular occupations. Recently, a population-based case–control study in Montreal found that workers with substantive exposure to asbestos had a greater risk of lung cancer, however, this finding did not achieve statistical significance (odds ratio (OR) =1.78, 95% CI=0.94, 3.36) [10]. Cumulative exposure was positively associated with lung cancer risk in a case–control study in Stockholm, Sweden [11], while a multi-center European case–control study found no association between occupational exposure to asbestos and lung cancer in six Central and Eastern European countries, but a nearly twofold (OR=1.85, 95% CI=1.07-3.21) increased risk was observed among UK workers [12].

While both cigarette smoking and asbestos are recognized lung carcinogens, there remain uncertainties about how they operate together to increase the risk of lung cancer. Attempts to understand the joint effects of smoking and asbestos on the risk of lung cancer extend back to Selikoff et al.'s seminal work in the late 1960s [13]. A subsequent review of this literature suggested that the interactive effects are multiplicative [14], which implies that asbestos exposure increases the risk of lung cancer by the same factor in smokers and non-smokers alike. An additive relationship, on the other hand, would assume that the effects of asbestos exposure and smoking are independent. Other reviews [15, 16] and a meta-analysis [17] have suggested that the combined effects of smoking and asbestos are more than additive but less than multiplicative. This conclusion is consistent with very recent work by Frost et al. that revealed interactions that were greater than additive, although the multiplicative association could not be rejected [18]. Apart from the studies by Gustavsson et al. and Pintos et al., we know of no other research that has evaluated the joint relationship between asbestos and smoking on lung cancer risk *in the general population* where exposure levels are much lower than in asbestos workers, yet with fewer precautions and protections offered to reduce exposure. In the Gustavsson et al. study, the association between asbestos and smoking on lung cancer risk was found to be between additivity and multiplicativity [11]. In the Montreal study, the association was found to be sub-multiplicative [10]. To add to this knowledge, we examined the joint relationship between smoking and asbestos in this population-based case–control study.

With this background, the primary objective of our study is to build upon past research by reporting on the association between occupational exposure to asbestos and lung cancer among Canadian men. The secondary objective of the study is to evaluate the combined effects occupational exposure to of asbestos and cigarette smoking on the risk of lung cancer.

METHODS

Study population

A case–control study design was used to address the research objectives, and the data come from the lung cancer case–control component of the National Enhanced Cancer Surveillance System

(NECSS). The overall objective of the NECSS was to improve our understanding of both environmental and occupational determinants of cancer [19]. The NECSS was a collaborative project between the Public Health Agency of Canada and cancer registries in eight Canadian provinces (British Columbia, Alberta,Saskatchewan, Manitoba, Ontario, Nova Scotia, Newfoundland, and Prince Edward Island). There were no subjects (cases or controls) from the province of Quebec. Detailed information was collected from cases and controls for a number of potential risk factors including: sociodemography, anthropometry, diet, smoking, exposure to second hand smoke, and participation in physical activities. Individuals were also asked to provide lifetime residential and occupational histories. Questionnaires were administered between 1994 and 1997.

The NECSS endeavoured to collect information for each incident cancer within three months of diagnosis. Among men, there were a total of 3,718 histologically confirmed lung cancer cases (ICD-9 rubric 162) identified between 1994 and 1997. Letters were sent to the physicians of 3,033 (81.6%) of these cases to solicit their participation. Physician consent was obtained and questionnaires were mailed to 2,548 (69%) of the cases; physician consent was refused for 229 (6%) of all eligible cases and 653 (18%) were deceased at the time of the request and therefore excluded. Completed questionnaires were returned by 1,736 of the 2,548 cases who were mailed a questionnaire yielding an overall response rate of 68.1%.

The NECSS assembled a series of controls from the general population. For 5 provinces, controls were identified through provincial health insurance plans (Prince Edward Island, Nova Scotia, Manitoba, Saskatchewan and British Columbia). These insurance plans cover more than 95% of residents in the province. Elsewhere, either random digit dialing (Newfoundland and Alberta), or property assessment data (Ontario) were used as the sampling frame to recruit controls. Frequency matching to the overall case grouping (19 types of cancers) was used to select controls with similar age and sex distribution, such that there would be at least one control for every case within each sex and 5-year age group for any specific cancer site within each province. In total, questionnaires were mailed to 4,270 men identified as possible controls in the 8 provinces. Approximately 7% of these (n=287) were returned because the address was incorrect, and no updated address could be found through publicly available sources. In all, 2,547 male controls returned completed questionnaires, representing 64% of those contacted and 60% of those ascertained.

For the purposes of our analyses, we restricted the study population to only include men given that we expected few women to have been exposed to asbestos in the workplace. We used the same analysis file previously used to evaluate associations between diesel engine exhaust emissions and lung cancer which excluded individuals under the age of 40, and those who had not worked for at least one year [20]. In the NECSS, among all participating incident lung cancer cases only 0.7% (n=13) were diagnosed before the age of 40; the corresponding number of controls excluded to meet the age requirement was 438. A total of 42 cases and 56 controls were excluded because their reported length of employment was less than one year. After applying these exclusion criteria we were left with a total of 1,681 cases and 2,053 controls.

Occupational assignment of exposures

Cases and controls were asked to provide information for each job held in Canada for at least 12 months from the time they were 18 years old until the time of interview. Information sought for each job included: job title, main tasks, type of industry, location, and the start and end dates of employment. A total of 15,646 jobs were identified, of these 15,234 (97.4%) jobs contained sufficient information for exposure assessment. No exposures were assigned for jobs that were self-reported to be retirement (n=185), disability (n=10), and unemployment (n=8).

Occupations and industry titles were assigned by one of two hygienists, who were blinded to case–control status, using the Canadian Classification and Dictionary of Occupation codes (originally published in 1971 with revisions up until 1986), and Standard Industrial Codes [21]. The hygienist coded each job on the basis of exposure to known or suspected lung cancer carcinogens. These exposures included: asbestos, diesel and gasoline engine exhaust emissions, and crystalline silica. This assessment was guided by the scientific and technical literature, consultation with experts, and a review of existing databases of exposure assessment. The assignment of workplace exposures took into account the manner that asbestos was used over the years. For example, before 1976, drywall installers used dry-wall joint cement that contained asbestos, while after 1980 asbestos was banned in this cement.

The assignment of occupational exposures was done according to three dimensions: concentration, frequency and reliability. The frequency of exposure was assigned based on the proportion of work time during a normal work week that the subject was exposed; this assignment took into account whether the work was part-time or seasonal in nature. *'Low'* frequency corresponded to less than 5% of the work time, *'Medium'* between 5% and 30%, and *'High'* represented more than 30%. Concentration was assessed on a relative scale. For each substance, benchmarks were established and exposures were coded with respect to these benchmarks. Non exposure was interpreted as exposure up to background levels found in the general environment. The relative benchmarks for concentration levels used by our team of hygienists were *'Low'* for welders and boiler operators, *'Medium'* for boiler and pipe insulators and marine firemen and *'High'* for miners and insulation workers (blowers and sprayers). It is very difficult to provide a reliable estimate of the absolute number of fibres per unit of volume corresponding to the different exposure levels. However, as a crude indicator, we can suggest that our 'Medium' level corresponded roughly to the 1976 American Conference of Governmental Industrial Hygienists threshold limit values (TLV) given that these values were in force in Canada in 1983 at a time when our study subjects were working. Specifically, the TLV for chrysotile asbestos fibers over 5 microns was 5 fibres per/cc in these Quebec guidelines. Finally the third dimension of exposure, reliability, refers to the hygienists' degree of confidence that the exposure was actually present in the job under evaluation; *'Low'* refers to a possible exposure, *'Medium'* to a probable exposure and *'High'* to a certain exposure. Estimates of the inter-rater reliability of the exposure assignment method, which were based on the work of chemists from the group that conducted the exposure assessment our study, lend credibility to the validity of the approach we used. Specifically, Goldberg et al. reported that the percent agreement among raters was between 95% to 98% with a Cohen's kappa from 0.5 to 0.7 [22].

Statistical analysis

We constructed several metrics to characterize occupational exposure to asbestos. These metrics included: ever exposed, highest attained concentration (high, medium, low), as well as a duration of exposure. Given the small number of individuals that had high concentrations of exposure, we combined medium and high into one group. Those with a low reliability score ("possibly exposed") were assumed to have had no exposure.

Logistic regression was used to estimate the odds ratios (OR) and their corresponding 95% confidence intervals (CI) for the various exposure metrics. Adjustments were made for the potential confounders: age, cigarette smoking, socioeconomic status, exposure to second hand smoke, and occupational exposure to silica, and diesel exhausts. Occupational exposure to silica, and diesel engine exhausts were assigned to the cases and controls using the same methodology that was used for asbestos. Silica and diesel exposures were modelled as cumulative time-weighted measures. While gasoline engine emission exposure measures were also derived for the cases and controls, they did not confound the risk estimates for asbestos, and therefore, were not included in the models as adjustment factors. Multivariable models were adjusted for cigarette smoking through the use of a pack-years variable which incorporated aspects of both smoking duration and intensity. Cigarette pack-years were defined as the number of years of smoking an average of 20 cigarettes per day. For exposure to second-hand smoke, a composite measure was used that took into account lifetime exposures received both at home, and in the workplace [23]. It was derived as a function of the number of years of exposure that incorporated both the number of regular smokers that lived in each residence, and the number of smokers who smoked regularly in the subjects' immediate work environment

The joint effect of smoking and occupational exposure to asbestos was first examined by estimating the odds ratios for cross-classification categories of cigarette pack-years (<10, 10 - <40, ≥40) and the highest attained occupational exposure to asbestos (none, low, medium/high). The small numbers of lung cancers among never smokers (n=34; 2% of all cases) precluded a separate evaluation of asbestos risks in this group. The odds ratios and 95% confidence intervals were estimated for eight cross-classification categories, while the ninth category (no asbestos exposure, < 10 cigarette pack-years) was used as the referent. The joint effects of smoking and asbestos on lung cancer risk were evaluated using two previously derived indices: the *Synergy* (S) [24] and *Multiplicativity* (V) [25]. We followed a similar approach that Frost et al. used to evaluate the relationship between asbestos and smoking and lung cancer in workers in Great Britain [18]. We used our derived odds ratios (ORs) to calculate the index S [24] as follows:

$$S = \frac{OR_{AS} - OR_0}{OR_A + OR_S - 2OR_0}$$

Where OR_A is the odds ratio of lung cancer exposed to 'medium or high' levels of asbestos among those with little to no smoking history (<15 pack-years), OR_S is the odds ratio of lung cancer among smokers (≥ 40 pack-years) with no exposure to asbestos, OR_{AS} is the odds ratio of lung cancer among smokers (≥ 40 pack-years) exposed to asbestos, where each odds ratio is estimated relative to the referent group of men who had accrued less than 10 cigarette pack-years and were not exposed to asbestos (OR_0).The Multiplicativity index was calculated as:

$$V = \frac{OR_0 OR_{AS}}{OR_A OR_S}$$

A value that exceeds one for the *S* index suggests an interactive effect between smoking and asbestos exposure on lung cancer that could imply a multiplicative effect. In contrast, a value of *S*near one suggests that the two risk factors would operate in an additive fashion on the risk of lung cancer. For the *V* index, a value of one indicates a multiplicative interaction, whereas as values greater and less than one indicate an interaction that is more or less than multiplicative, respectively.

Ethics approval

The participating provincial cancer registries obtained approval of the NECSS study protocol through their respective ethics review boards. All participants provided informed consent.

RESULTS

Of the 15,234 occupations ever held by the study subjects, a total of 801 were coded as having either 'probable' or 'definite' exposure to asbestos. The most commonly reported exposed occupations were mechanics and repairmen, stationary engine and utility workers, pipefitters, and construction workers (Table 1). Water transport operating occupations represented the only group deemed to have a high frequency of exposure to asbestos. Specific jobs included in this group that worked on ships included: deck officers, engineering officers, deck crew, engine and boiler room crew workers.

Table 1. Most frequent occupations among the 801 jobs held by subjects that were classified as having probable or definite exposure to asbestos

	SOC	Number of jobs	%	Most common exposure coding[A]		
				Confidence	Frequency	Concentration
Mechanics and Repairmen (except electrical)	8580 – 8589	214	26.7	Probable	Low	Low
Stationary Engine and Utilities Equipment	9530 - 9539	124	15.0	Probable	Medium	Low
Pipefitting	8791	89	11.1	Probable	Low	Low
Construction	8733	79	9.9	Probable	Low	Low
Metal shaping occupations	8330 – 8339	48	6.0	Probable	Medium	Low
Fabricating, assembling electrical and electronics	8530 – 8539	34	4.2	Probable	Medium	Low
Water Transport Operating Occupations	9151 – 9159	42	5.2	Probable	High	Low
Firefighters	6111	31	3.9	Definite	Medium	Low
Plasterers	8784	23	2.9	Probable	Medium	Low
Total		801	84.5			

A – defined by highest percentage.

Table 2. Adjusted odds ratios of lung cancer in relation to occupational exposure to asbestos

Occupational exposure	Cases	Controls	Odds ratio[A] and 95% CI		Odds ratio[B] and 95% CI	
Unexposed*	1448	1829	1.0	-	1.0	-
Ever exposed	233	224	1.31	1.07 – 1.59	1.28	1.02 – 1.61
Highest attained exposure						
Unexposed	1448	1829	1.0	-	1.0	
Low	194	200	1.22	0.99 – 1.51	1.17	0.92 – 1.50
Medium / High	39	24	2.02	1.20 – 3.97	2.16	1.21 – 3.88
Total	1681	2053				
Duration of exposure (years)						
< 10	88	68	1.68	1.21 – 2.33	1.60	1.10 – 2.33
10 - < 20	46	50	1.08	0.74 – 1.69	0.89	0.56 – 1.42
≥ 20	87	103	1.05	0.78 – 1.42	1.18	0.84 – 1.66

A – Adjusted for age, province,

B – Adjusted for age, province, cigarette pack years, occupational exposure to diesel and silica, exposure to second hand smoke.

A total of 233 cases and 224 controls, respectively, were exposed to asbestos at some point during their lifetime occupational history (Table 2). Those who were ever exposed to asbestos had a 28%

increased risk of lung cancer relative to those who were not (OR=1.28, 95% CI: 1.02, 1.61). The risks according to highest concentration of occupational exposure ever attained were more pronounced. Only two cases and one control reported working in a job with an assigned 'high' concentration of exposure. As a result, we combined 'medium' and 'high' concentrations into one category. Those who had ever been exposed to medium or high levels had a more than twofold increase in risk (OR=2.16, 95% CI=1.21-3.88).

We found that duration of occupational exposure to asbestos was not related to the risk of lung cancer (Table 2). When we modeled duration of exposure as a continuous variable, the adjusted odds ratio of lung cancer for an increase in 10 years of exposure was 1.03 (95%% CI=0.94-1.13). This risk increased to 1.13 (95% CI=0.84-1.52) when analyses were restricted to those who were only exposed to medium or high concentrations; this result however was not statistically significant (p=0.44). The frequency of the jobs that were deemed to have 'medium' or 'high' concentrations of asbestos is presented in Figure 1. The most common of these jobs were pipefitters and boilermakers, and insulators.

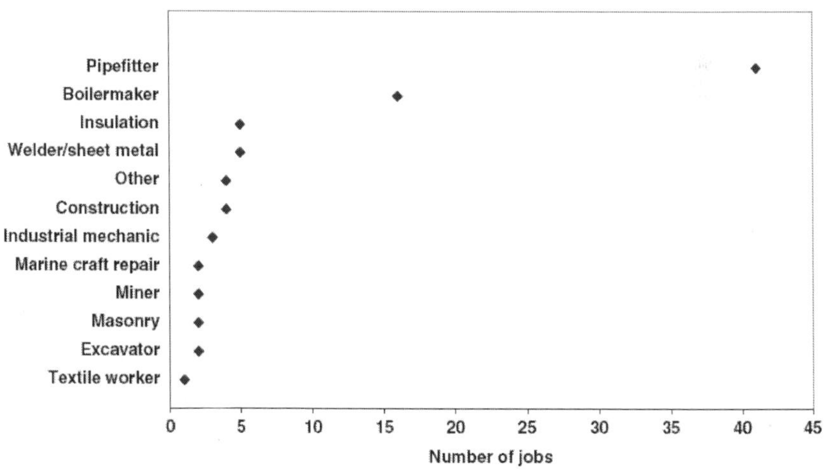

Figure 1. Most common occupations among mean with medium or high concentration levels of asbestos, NECSS lung cancer case-control study.

None of the first-order interaction terms between cigarette smoking pack-years and the three measures of asbestos exposure were statistically significant. The corresponding p-values for the smoking interaction terms with ‘ever’, ‘highest attained’ and ‘duration’ asbestos exposure were 0.33, 0.77, and 0.88, respectively.

Table 3. Adjusted odds ratios* and 95% C.I. according highest occupational exposure to asbestos across cigarette pack-year smoking categories

	Cigarette smoking (pack-years)								
Highest occupational exposure to asbestos	< 10			10 - <40			≥ 40		
	N	OR*	95% C.I.	N	OR*	95% C.I.	N	OR*	95% C.I.
None	84	1.0	-	630	1.0	-	678	1.0	-
Low	13	1.54	0.79 – 3.00	90	1.28	0.91 – 1.80	85	0.88	0.58 – 1.35
Medium or high	2	2.01	0.39 – 10.43	20	2.30	1.09 – 4.84	16	2.50	0.73 – 9.28

N = number of lung cancer cases.

* adjusted for age, province, occupational exposure to diesel and silica, and second hand cigarette smoke.

Table 4. Synergy and multiplicative indices between asbestos exposure and cigarette smoking

Cigarette smoking (pack-years)	Asbestos exposure	Label	Cases	Controls	Odds ratio*	95% CI
< 10	None	R_0	84	745	1.0	-
	Low	—	13	69	1.47	(0.77 – 2.81)
	Medium/High	R_A	2	7	2.20	0.42 – 11.41)
10 - < 40	None	—	630	778	5.28	(3.90 – 7.14)
	Low	—	90	90	6.67	(4.41 – 10.10)
	Medium/High	—	20	13	10.39	(4.83 – 22.36)
≥ 40	None	R_S	678	266	17.68	(12.90 – 24.22)
	Low	—	85	40	15.62	(9.72 – 25.09)
	Medium/High	R_{AS}	16	3	38.59	(10.78 – 138.08)
Synergy Index					2.10	
Multiplicativity Index					0.99	

* adjusted for age, province, occupational exposure to diesel and silica, and second hand cigarette smoke.

Stratified analyses of highest attained asbestos exposure across cigarette pack years categories are presented in Table 3. There was an approximate two-fold increase in risk among those with ‘medium’ or ‘high’ occupational exposure to asbestos relative to those with no such exposure in each of the three pack-year categories. This is consistent with a multiplicative relationship between the two factors. Those who had at least 40 pack-years of smoking and were exposed to medium or high asbestos levels had the highest risk of lung cancer; relative to those with no asbestos exposure, and less than 10 cigarette pack-years, their risk nearly 38-fold higher (OR=38.59, 95% CI=10.78-138.08) (Table 4). The calculated values of the S and V indices were 2.10 and 0.99 respectively, supporting the notion that the interaction between asbestos and smoking is multiplicative.

DISCUSSION

This population-based study of men employed across a diverse range of jobs found that workplace exposure to asbestos was associated with an increased risk of lung cancer. This association persisted after adjusting for cigarette smoking, second hand smoke, and other occupational exposures previously implicated as possible risk factors for lung cancer. The approximate 28% increased risk observed among men ever exposed to asbestos is similar to the finding of Pintos et al. [10]. In their Montreal based case–control study, those who were exposed to asbestos had an odds ratio of 1.21, (95% CI=0.98-1.49) relative to those with no exposures. The population attributable risk (PAR) percent is often used to provide an estimate of the percentage of cases that be avoided if the putative exposure was eliminated [26]. We calculated the PAR in our study using the odds ratio of 1.28 among ever exposed, and an estimated prevalence of exposure of 11.3% (based on our control series). This yielded a PAR of 3.1% which suggests that a relatively small percentage of Canadian male lung cancer cases are due to occupational exposure to asbestos. Based on an estimated 13,300 incident lung cancers among men in Canada in 2012 [1] this would account for approximately 412 incident cases.

Our study provided support for a dose–response relationship between asbestos exposure and lung cancer as higher risks were observed among those who were ever exposed to 'medium' or 'high' concentrations of asbestos. Pipefitters accounted for nearly half of these cases and controls (41 of 87). While the limited number of subjects did not allow us to characterize risks for specific types of jobs, our results are consistent with a previously published study of Ontario pipe trade workers [27]. They reported a 53% increased risk of lung cancer mortality among pipefitters who had been registered trade members for at least 30 years, relative to the Ontario general population. However, their study was somewhat limited due to a lack of data on smoking. Our findings support the hypothesis that asbestos and cigarette smoking affect the risk of lung cancer in a multiplicative fashion.

In many occupational studies, duration of exposure is regarded as valid surrogate measure of cumulative exposure due to the inherent difficulties in retrospective studies to precisely characterize exposure intensity. In their Montreal case–control study, Pintos et al. found a higher risk of lung cancer among those exposed to asbestos for at least 20 years when compared to those exposed for shorter durations [10]. Duration of exposure was also positively associated with lung cancer risk in other industry-specific cohorts [28]. In contrast, we found that only intensity but not duration of exposure was associated with statistically significant increased risks of lung cancer. This observation is consistent with recently published findings on a cohort of workers employed in an asbestos reprocessing plant in the Calvados region of France [29]. In this study, Clin and colleagues observed that the average exposure to asbestos expressed in terms of fibers per ml was associated with pleuro-peritoneal mesothelioma, lung cancer, and colorectal cancer ($p<0.05$), however, no statistically significant associations were evident with duration of exposure for any of these three cancer sites. Other studies of asbestos workers have also found associations with intensity but not duration of exposure [12, 30, 31]. Our finding of a stronger positive association between duration of exposure at medium or high levels of asbestos when compared to durations spent at lower levels suggests that time exposed above a threshold level may be a relevant marker of risk. However, this finding should be interpreted cautiously as it based on a very small number of subjects who were exposed to either medium or high intensities.

It is well recognized that there is a lengthy latency period between the time of first exposure to an environmental carcinogen and the development of a solid tumour such as lung cancer. For example, the latency period associated with cigarette smoking and lung cancer has been estimated to be several decades following the initiation of smoking [32]. By extension, the increased risks of lung cancer due to exposure to asbestos observed in this study are a reflection of workplace exposures many years if not decades earlier. Indeed, among those classified has having 'medium' or 'high' concentrations to asbestos in the workplace, the start date of employment was after 1980 in only 6% of these jobs.

Participants in our study were asked to provide information for only those jobs that were held for at least one year. The exclusion of these short-term jobs raises the possibility that some exposure misclassification has been introduced. Previous analysis of 27.5 million workers found increased risks of lung cancer among those exposed to high levels of asbestos (20 to 40 fibers per cubic centimeter of air) for only a few months [33]. Under a classical error model where the possible exposure misclassification error arising from excluding these short term jobs is non-differential to case–control status, our risk estimates would be understated.

An important strength of this study was the availability of other risk factor data obtained through both the questionnaire, as well as expert-based coding of occupational histories. Unlike many other occupational case–control studies, we had extensive data on cigarette smoking, most notably, exposure to second hand smoke. This measure allowed our risk estimates to take into account lifetime exposure to second-hand smoke incurred at both home and workplace settings. In addition, the industrial hygienists also coded each job for possible exposure to other known or suspected lung carcinogens including: crystalline silica, gasoline and engine emissions. We recently found that occupational exposure to diesel but not gasoline engine emissions increased the risk of lung cancer; the risk of lung cancer was also increased among individuals exposed to crystalline silica [34]. The addition of these two covariates (diesel and silica) strengthened the association for asbestos by approximately 20%.

Approximately 68% of eligible cases and 64% of eligible controls completed a questionnaire. This raises the potential to introduce some bias in our risk estimates, and our results should be interpreted cautiously because of this possibility. However, for several reasons, we do not believe this bias fundamentally changes our results. First, observed associations with known and suspected risk factors such as cigarette smoking, and exposure to second-hand smoke are similar in direction and magnitude to risk estimates reported in other epidemiological studies. Moreover, our published findings for other occupational exposures within the same study population [34] are also consistent with the epidemiological literature. Lastly, the distribution of lung cancers by histology in our study is remarkably similar to population-based figures for North America [35] and provides some support for the generalizeability of these results to incident lung cancers in Canada. Unfortunately, the NECSS did not collect data from those diagnosed with mesothelioma, and therefore, we were unable to investigate associations with this endpoint.

We were unable to distinguish asbestos on the basis of fiber type. Asbestos fibers can be described according to two broad classes *serpentines* (phyllosilicates) and *amphiboles* (inosilicates) that differ substantially with respect to biopersistence and physical and chemical properties. Serpentines include chrysotile asbestos which is the predominant type of asbestos in Canada. The International Agency for Research on Cancer has determined that there is sufficient evidence to conclude that all these forms of asbestos can cause cancer in humans [4, 6]. There remains considerable uncertainty regarding differences in lung cancer risk resulting from exposure to different types of asbestos fibers. A review of cohort studies where quantitative measurements of asbestos exposure were available demonstrated clearer and consistent associations between exposure and lung cancer for crocidolite or amosite [36]. On the other hand, associations from cohorts exposed primarily to crysotile asbestos were less consistent [37, 38]. It is generally accepted that amphibole fibers are more harmful than chrysotile fibers for mesothelioma [36, 39]. However, it has been argued that these differences are not all that important given that chrysotile is the most commonly used type of asbestos [40, 41]. In our study, those who were determined to have been exposed to asbestos were believed to have been exposed to chrysotile, however, it is possible that some exposure to less prevalent yet more potent types of fibers occurred and was unaccounted for.

Another limitation of our study was the relatively small number of study subjects who were ever exposed to medium or high levels of asbestos. In total, there were only 39 cases and 24 controls exposed at these levels. These small numbers hindered our ability to characterize the joint relationship between smoking and asbestos exposure on the risk of lung cancer. It also limited our examination of the risks of lung cancer with exposure to asbestos according to different histological subtypes. Several studies have found associations that were most pronounced for adenocarcinoma subtypes [28, 42–44], however, others did not [45–47]. The three most common histological types of lung cancer in our study population were squamous cell carcinoma (35%), adenocarcinoma (28%), and small cell carcinoma (15.9%) [34]. When we restricted analysis to adenocarcinoma, the odds ratio among those exposed to medium or high levels of asbestos increased from 2.16 to 3.14 (95% CI=1.50 – 6.58). However, the latter estimate was based on only 13 incident cases and therefore, our study has very limited statistical power to make inferences by histological type.

CONCLUSIONS

In summary, the findings from this Canadian case–control study are consistent with the determination by international agencies that asbestos is a human lung carcinogen. While chrysotile asbestos is the predominant type of asbestos in Canada, it is possible that some of the workers in our study were exposed to other types of asbestos fibers. For this reason, and given the relatively small number of individuals exposed to medium and high exposure where the excess risks of lung cancer were found, we cannot conclusively attribute increased lung cancer risks to chrysotile. Despite the limitation, our findings provide further support that exposure to asbestos has contributed to an increased risk of lung cancer in Canadian workplaces.

AUTHORS' INFORMATION

The Canadian Cancer Registries Epidemiology Research Group comprised a principal investigator from each of the provincial cancer registries involved in the National Enhanced Cancer Surveillance System: Bertha Paulse, Newfoundland Cancer Foundation; Ron Dewar, Nova Scotia Cancer Registry; Dagny Dryer, Prince Edward Island Cancer Registry; Nancy Kreiger, Cancer Care Ontario; Erich Kliewer, Cancer Care Manitoba; Diane Robson, Saskatchewan Cancer Foundation; Shirley Fincham, Division of Epidemiology, Prevention and Screening, Alberta Cancer Board; and Nhu Le, British Columbia Cancer Agency.

ACKNOWLEDGEMENTS

We thank Benoit Latreille and Louise Nadon for their tireless efforts in assigning the occupational exposures. We are also grateful to the helpful comments provided on earlier drafts of this manuscript by Paul Demers of the Ontario Occupational Cancer Research Centre, and Michel Camus of Health Canada. This study was funded by Health Canada.

AUTHORS' CONTRIBUTIONS

PV contributed to the design of the study, conducted analysis of the data, and took the lead in preparing the manuscript. SH contributed to the design of the study, and assisted in the development of the manuscript. KJ is the principle investigator of the NECSS and oversaw the original collection of the data, design of the questionnaire, and contributed to the writing of this manuscript. MEP oversaw the assignment of the occupational exposures for this study, contributed to the design, and played a prominent role in the writing of this manuscript. All authors read and approved the final manuscript.

REFERENCES

1. Canadian Cancer Society's Steering Committee: Canadian Cancer Statistics 2012. 2012, Toronto: Canadian Cancer Society
2. Vitra RL: Worldwide Asbestos Supply and Consumption Trends from 1900 through 2003. 2006, Reston, Virginia: US Geological Survey Circular 1298, 80-
3. United States Environmental Protection Agency: Airborne Asbestos Health Assessment Update. 1986, Washington, DC: US EPA
4. International Agency for Research on Cancer: IARC Monographs in the Evaluation of the Carcinogenic Risk of Chemicals to Humans. Overall Evaluations of Carcinogenicity: An updating of IARC Monographs to Humans Volumes 1 to 42. 1987, Lyon, France: IARC
5. National Toxicology Program: Report on carcinogens. 1980, Washington, DC: US Department of Health and Human Services, 1
6. International Agency for Research on Cancer: IARC Monographs on the Evaluation of the Carcinogenic Risk of Chemicals to Man, vol 14 Asbestos. 1977, Lyon, France: IARC
7. Ramazzini C: Asbestos is still with us: repeat call for a universal ban. J Occup Environ Med. 2010, 52 (5): 469-472.

8. Asbestos: elimination of asbestos-related diseases. Fact Sheet Number 343. [http://www.who.int/mediacentre/factsheets/fs343/en/index.html],

9. Asbestos mining stops for first time in 130 years. [http://www.cbc.ca/news/canada/story/2011/11/24/asbestos-shutdown.html],

10. Pintos J, Parent ME, Rousseau MC, Case BW, Siemiatycki J: Occupational exposure to asbestos and man-made vitreous fibers, and risk of lung cancer: evidence from two case–control studies in Montreal, Canada. J Occup Environ Med. 2008, 50 (11): 1273-1281.

11. Gustavsson P, Nyberg F, Pershagen G, Scheele P, Jakobsson R, Plato N: Low-dose exposure to asbestos and lung cancer: dose–response relations and interaction with smoking in a population-based case-referent study in Stockholm, Sweden. Am J Epidemiol. 2002, 155 (11): 1016-1022.

12. Carel R, Olsson AC, Zaridze D, Szeszenia-Dabrowska N, Rudnai P, Lissowska J, Fabianova E, Cassidy A, Mates D, Bencko V, et al: Occupational exposure to asbestos and man-made vitreous fibres and risk of lung cancer: a multicentre case–control study in Europe. Occup Environ Med. 2007, 64 (8): 502-508.

13. Selikoff IJ, Hammond EC, Churg J: Asbestos exposure, smoking, and neoplasia. JAMA. 1968, 204 (2): 106-112.

14. Vainio H, Boffetta P: Mechanisms of the combined effect of asbestos and smoking in the etiology of lung cancer. Scand J Work Environ Health. 1994, 20 (4): 235-242.

15. Berry G, Liddell FD: The interaction of asbestos and smoking in lung cancer: a modified measure of effect. Ann Occup Hyg. 2004, 48 (5): 459-462.

16. Liddell FD: The interaction of asbestos and smoking in lung cancer. Ann Occup Hyg. 2001, 45 (5): 341-356.

17. Wraith D, Mengersen K: Assessing the combined effect of asbestos exposure and smoking on lung cancer: a Bayesian approach. Stat Med. 2007, 26 (5): 1150-1169.

18. Frost G, Darnton A, Harding AH: The effect of smoking on the risk of lung cancer mortality for asbestos workers in Great Britain (1971–2005). Ann Occup Hyg. 2011, 55 (3): 239-247.

19. Johnson KC, Mao Y, Argo J, Dubois S, Semenciw R, Lava J: The National Enhanced Cancer Surveillance System: a case–control approach to environment-related cancer surveillance in Canada. Envirometrics. 1998, 9: 495-504.

20. Villeneuve PJ, Parent ME, Sahni V, Johnson KC, Canadian Cancer Registries Epidemiology Research G: Occupational exposure to diesel and gasoline emissions and lung cancer in Canadian men. Environ Res. 2011, 111 (5): 727-735.

21. Department of Manpower and Immigration, Immigration Canada: Canadian classification and dictionary of occupations. Vol 1. 1971, Ottawa, Canada: Immigration Canada

22. Goldberg MS, Siemiatycki J, Gerin M: Inter-rater agreement in assessing occupational exposure in a case–control study. Br J Ind Med. 1986, 43 (10): 667-676.

23. Johnson KC, Hu J, Mao Y: Passive and active smoking and breast cancer risk in Canada, 1994–97. Cancer Causes Control. 2000, 11 (3): 211-221.

24. Rothman KJ: The estimation of synergy or antagonism. Am J Epidemiol. 1976, 103 (5): 506-511.

25. Lee PN: Relation between exposure to asbestos and smoking jointly and the risk of lung cancer. Occup Environ Med. 2001, 58 (3): 145-153.

26. Last JM: A Dictionary of Epidemiology. 2000, New York: Oxford University Press, 4

27. Finkelstein MM, Verma DK: A cohort study of mortality among Ontario pipe trades workers. Occup Environ Med. 2004, 61 (9): 736-742.

28. Raffn E, Lynge E, Korsgaard B: Incidence of lung cancer by histological type among asbestos cement workers in Denmark. Br J Ind Med. 1993, 50 (1): 85-89.

29. Clin B, Morlais F, Launoy G, Guizard AV, Dubois B, Bouvier V, Desoubeaux N, Marquignon MF, Raffaelli C, Paris C, et al:

Cancer incidence within a cohort occupationally exposed to asbestos: a study of dose–response relationships. Occup Environ Med. 2011, 68 (11): 832-836.

30. Yano E, Wang X, Wang M, Qiu H, Wang Z: Lung cancer mortality from exposure to chrysotile asbestos and smoking: a case–control study within a cohort in China. Occup Environ Med. 2010, 67 (12): 867-871.

31. Gustavsson P, Jakobsson R, Nyberg F, Pershagen G, Jarup L, Scheele P: Occupational exposure and lung cancer risk: a population-based case-referent study in Sweden. Am J Epidemiol. 2000, 152 (1): 32-40.

32. Bilello KS, Murin S, Matthay RA: Epidemiology, etiology, and prevention of lung cancer. Clin Chest Med. 2002, 23 (1): 1-25.

33. Seidman H, Selikoff IJ, Gelb SK: Mortality experience of amosite asbestos factory workers: dose–response relationships 5 to 40 years after onset of short-term work exposure. Am J Ind Med. 1986, 10 (5–6): 479-514.

34. Villeneuve PJ, Parent ME, Sahni V, Johnson KC: Occupational exposure to diesel and gasoline emissions and lung cancer in Canadian men. Environ Res. 2011

35. Wynder EL, Muscat JE: The changing epidemiology of smoking and lung cancer histology. Environ Health Perspect. 1995, 103 (Suppl 8): 143-148.

36. Hodgson JT, Darnton A: The quantitative risks of mesothelioma and lung cancer in relation to asbestos exposure. Ann Occup Hyg. 2000, 44 (8): 565-601.

37. Dement JM, Brown DP, Okun A: Follow-up study of chrysotile asbestos textile workers: cohort mortality and case–control analyses. Am J Ind Med. 1994, 26 (4): 431-447.

38. Liddell FD, McDonald AD, McDonald JC: The 1891–1920 birth cohort of Quebec chrysotile miners and millers: development from 1904 and mortality to 1992. Ann Occup Hyg. 1997, 41 (1): 13-36.

39. Berman DW, Crump KS: Update of potency factors for asbestos-related lung cancer and mesothelioma. Crit Rev Toxicol. 2008, 38 (Suppl 1): 1-47.

40. Stayner LT, Dankovic DA, Lemen RA: Occupational exposure to chrysotile asbestos and cancer risk: a review of the amphibole hypothesis. Am J Public Health. 1996, 86 (2): 179-186.

41. Cullen MR: Chrysotile asbestos: enough is enough. Lancet. 1998, 351 (9113): 1377-1378.

42. Hourihane DO, McCaughey WT: Pathological aspects of asbestosis. Postgrad Med J. 1966, 42 (492): 613-622.

43. Johansson L, Albin M, Jakobsson K, Mikoczy Z: Histological type of lung carcinoma in asbestos cement workers and matched controls. Br J Ind Med. 1992, 49 (9): 626-630.

44. Whitwell F, Newhouse ML, Bennett DR: A study of the histological cell types of lung cancer in workers suffering from asbestosis in the United Kingdom. Br J Ind Med. 1974, 31 (4): 298-303.

45. Kannerstein M, Churg J: Pathology of carcinoma of the lung associated with asbestos exposure. Cancer. 1972, 30 (1): 14-21.

46. Lee BW, Wain JC, Kelsey KT, Wiencke JK, Christiani DC: Association of cigarette smoking and asbestos exposure with location and histology of lung cancer. Am J Respir Crit Care Med. 1998, 157 (3 Pt 1): 748-755.

47. Warnock ML, Isenberg W: Asbestos burden and the pathology of lung cancer. Chest. 1986, 89 (1): 20-26.

Chapter 8

Experimental Exposure to Diesel Exhaust Increases Arterial Stiffness In Man

Magnus Lundbäck*[1], Nicholas L Mills[2], Andrew Lucking[2], Stefan Barath [1], Ken Donaldson[3], David E Newby[2], Thomas Sandström[1] and Anders Blomberg[1]

[1]*Department of Respiratory Medicine and Allergy, University Hospital, Umeå, Sweden,*

[2]*Centre for Cardiovascular Science, Edinburgh University, Edinburgh, UK*

[3]*ELEGI Colt Laboratory, Centre for Inflammation Research, Edinburgh University, Edinburgh, UK*

ABSTRACT

Introduction: Exposure to air pollution is associated with increased cardiovascular morbidity, although the underlying mechanisms are unclear. Vascular dysfunction reduces arterial compliance and increases central arterial pressure and left ventricular after-load. We determined the effect of diesel exhaust exposure on arterial compliance using a validated non-invasive measure of arterial stiffness.

Methods: In a double-blind randomized fashion, 12 healthy volunteers were exposed to diesel exhaust (approximately 350 μg/m^3) or filtered air for one hour during moderate exercise. Arterial stiffness was measured using applanation tonometry at the radial artery for pulse wave analysis (PWA), as well as at the femoral and carotid arteries for pulse wave velocity (PWV). PWA was performed 10, 20 and 30 min, and carotid-femoral PWV 40 min, post-exposure. Augmentation pressure (AP), augmentation index (AIx) and time to wave reflection (Tr) were calculated.

Results: Blood pressure, AP and AIx were generally low reflecting compliant arteries. In comparison to filtered air, diesel exhaust exposure induced an increase in AP of 2.5 mmHg ($p = 0.02$) and in AIx of 7.8% ($p = 0.01$), along with a 16 ms reduction in Tr ($p = 0.03$), 10 minutes post-exposure.

Conclusion: Acute exposure to diesel exhaust is associated with an immediate and transient increase in arterial stiffness. This may, in part, explain the increased risk for cardiovascular disease associated with air pollution exposure. If our findings are confirmed in larger cohorts of susceptible populations, this simple non-invasive method of assessing arterial stiffness may become a useful technique in measuring the impact of real world exposures to combustion derived-air pollution.

BACKGROUND

The adverse health effects of air pollution have been studied intensely over the last 50 years. Epidemiological studies have identified a strong association between air pollution and increased cardiovascular and respiratory morbidity and mortality [1]. Airborne particles with a diameter of less than 2.5 μm ($PM_{2.5}$) are considered the main cause of these adverse health effects [2]. In most countries, diesel exhaust (DE) is a major contributor to combustion-derived PM in urban areas [3], exposure to which appears to be an important trigger for acute myocardial infarction [4].

In recent years, there has been considerable interest and scientific research into the mechanisms that underpin these associations. We have recently demonstrated that exposure to DE causes exercise-induced ST-segment depression and impairs endogenous fibrinolytic function in patients with stable coronary heart disease [5]. Brook and colleagues suggest that exposure to air pollution may result in acute arterial vasoconstriction detected by ultrasonography of the brachial artery [6]. In complementary studies using venous occlusion plethysmography, we have reported impaired vasodilatation in response to intra-arterial infusion of endothelium-dependent and -independent agonists immediately after exposure to DE [7]. These sophisticated techniques are ideal for small scale experimental exposure studies. However, there is a need to develop a simple non invasive bedside test to assess the effects of real world exposures to ambient air pollution on vascular function in larger population studies.

The vascular endothelium plays a key role in the regulation of arterial tone. Traditional cardiovascular risk factors are associated with vascular dysfunction, reduced arterial compliance and increased central aortic pressure [8]. Enhanced large artery stiffness results in greater aortic systolic pressure, increased left ventricular afterload and reduced coronary blood flow [9]. Arterial stiffness may be the central pathological process in essential hypertension and is an independent predictor of mortality [10]. Applanation tonometry of the radial artery with pulse wave analysis is a well validated non invasive method of assessing arterial stiffness [11, 12]. Smokers have been found to have stiffer arteries compared to non-smokers and develop an additional increase in arterial stiffness shortly after smoking one cigarette [13]. Given the similarity between cigarette smoke and combustion-derived air pollution, we hypothesised that exposure to DE would alter arterial tone and stiffness. The aim of this study was to assess the effects of exposure to diesel exhaust on arterial stiffness and blood pressure in healthy non-smoking volunteers.

METHODS

Subjects

Twelve, non smoking, healthy men (mean age 26, range 21–30) participated in the study. All subjects had normal lung function, and reported no symptoms of respiratory tract infection for at least six weeks before or during the study. Exclusion criteria were regular medication, clinical evidence of atherosclerotic vascular disease, arrhythmias, diabetes mellitus, hypertension, renal or hepatic failure, asthma, occupational exposure to air pollution or an intercurrent illness. These subjects also participated in a separate study protocol that examined the effects of diesel exhaust exposure on thrombus formation and platelet activation; the results of that study are reported separately [14]. The study was approved by the local ethics committee, performed in accordance to the Declaration of Helsinki, and with the written informed consent of all participants.

Study design

In a randomised double blind crossover study, subjects attended on two occasions at least one week apart and were exposed to either diesel exhaust or filtered air for 60 min. On each occasion, subjects underwent moderate exercise on a bicycle ergometer alternated with rest at 15 minutes intervals throughout, according to a previously described standard protocol [7]. The subjects abstained from alcohol for 24 hours and from food, tobacco and caffeine containing drinks for at least 4 hours before exposure. Immediately after exposure, arterial stiffness was performed with measures of pulse wave analysis (PWA) at 10, 20 and 30 minutes and pulse wave velocity (PWV) at 40 minutes after the end of exposure.

Diesel-exhaust exposure

Temperature and humidity in the chamber were controlled at 22°C and 50% respectively.

The diesel exhaust was generated from an idling Volvo diesel engine (Volvo, TD45, 4.5 L, 4 Cylinders, 680 rpm) using Gasoil E10 (Preem, Sweden). After more than 90% of the exhaust was shunted away, the remainder was diluted with filtered air and fed into the exposure chamber at a steady-state concentration. Air in the chamber was continuously monitored for pollutants with exposures standardised using continuous measurement of oxides of nitrogen concentrations (NO_x) to deliver a particulate concentration of approximately 350 μg/m^3. Exposures were carried out in accordance to a well established protocol as previously reported. Workload during both exposures was adjusted to achieve the preset minute ventilation of 25 L/min/m^2. We have previously ascertained by a questionnaire that there are no differences in reported symptoms (including those related to the upper airways) between the diesel and filtered air exposures and volunteers are unable to tell which exposure they have received. [15].

Arterial Stiffness

Vascular studies were performed in a quiet, temperature controlled room with subjects resting in a supine position. Systolic and diastolic blood pressures were measured in duplicate using a semi automated non-invasive oscillometric sphygmomanometer, following a 10 min rest period and at 10, 20 and 30 minutes after each exposure.

Pulse wave analysis was performed using micromanometer applanation tonometry (Millar Instruments, Texas) of the radial artery at the wrist and the SphygmoCor™ system (AtCor Medical, Sydney) in accordance with the manufacturer's recommendations. Briefly, pulse wave analysis derives an aortic pulse pressure waveform from the radial artery wave via a mathematical transfer function. The arterial pressure waveform is a composite of the forward pressure wave created by ventricular contraction and a reflected wave generated by peripheral vascular resistance. The augmentation pressure (AP), is the difference between the second and first systolic peaks expressed in mm(Hg). The augmentation index (AIx), augmentation pressure as a percentage of the pulse pressure, is a measure of systemic arterial stiffness and wave reflection. The time to wave reflection (Tr) is reduced with increasing arterial stiffness, and provides a surrogate of aortic pulse wave velocity [16]. At least two independent waveform analyses were obtained from each subject, with measurements only accepted upon meeting SphygmoCor™ quality control criteria. Pulse wave velocity (PWV) was calculated by measuring the time for the pulse wave to travel between the carotid and femoral arteries. All measurements were performed by a single operator blinded to the nature of each exposure.

Statistical analysis

The results were analysed using GraphPad Prism (GraphPad Software, USA) by 2-tailed Student's *t*-test and a repeated measure ANOVA where appropriate. Statistical significance was taken at $P < 0.05$. Data are presented as mean ± SEM.

RESULTS

Between DE exposures there were small variations in particle mass (330 ± 12 $\mu g/m^3$), estimated particle number ($1.26 \pm 0.01 \times 10^6/cm^3$), oxides of nitrogen, NO_x (2.78 ± 0.03 ppm), nitrogen dioxide (0.62 ± 0.01 ppm), nitric oxide (2.15 ± 0.03 ppm), carbon monoxide (3.08 ± 0.12 ppm) and total hydrocarbons (1.58 ± 0.16 $\mu g/m^3$).

Table 1. Effects of diesel exposure on heart rate, blood pressure and markers of arterial stiffness

	Air (n = 12)			Diesel exhaust (n = 12)			Significance
Time, mins	+10	+20	+30	+10	+20	+30	-
Heart rate, bpm	57 ± 2	55 ± 2	55 ± 2	57 ± 2	55 ± 2	54 ± 2	P = 0.66
Systolic BP, mmHg	118 ± 2	114 ± 2	115 ± 3	114 ± 2	114 ± 2	115 ± 3	P = 0.26
Diastolic BP, mmHg	67 ± 2	65 ± 2	66 ± 2	65 ± 2	68 ± 2	67 ± 3	P = 0.48
MAP, mmHg	80 ± 2	78 ± 2	79 ± 2	79 ± 2	81 ± 3	81 ± 3	P = 0.57
Augmentation pressure, mmHg	-2.3 ± 0.7	-2.2 ± 0.8	-2.5 ± 0.7	0.2 ± 0.9	-1.3 ± 0.9	-1.8 ± 0.8	P = 0.01
Augmentation index, %	-7.0 ± 2.3	-7.4 ± 2.6	-7.9 ± 2.2	0.8 ± 3.2	-3.8 ± 3.5	-5.8 ± 2.7	P = 0.02
Time to wave reflection, ms	187 ± 8	189 ± 11	180 ± 7	171 ± 11	163 ± 9	180 ± 8	P = 0.03

Values are Mean ± SEM;

diesel exhaust *versus* filtered air (repeated measures ANOVA)

BP = blood pressure; MAP = mean arterial pressure

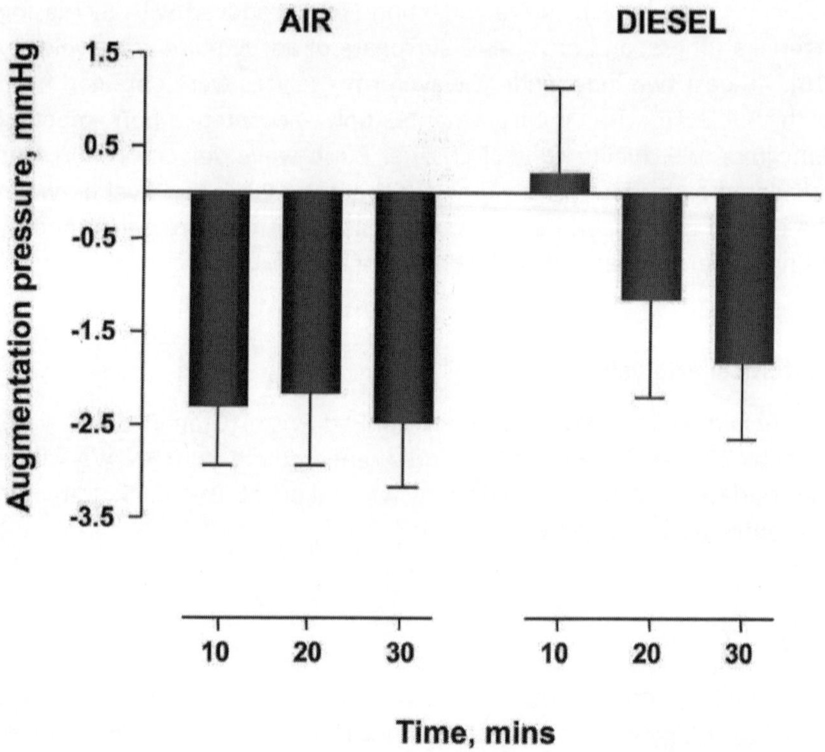

Figure 1. Augmentation pressure increases immediately following exposure to dilute diesel exhaust as compared to filtered air and is normalised over 30 minutes (P = 0.01, diesel exhaust ***versus*** filtered air, repeated measures ANOVA, n = 12).

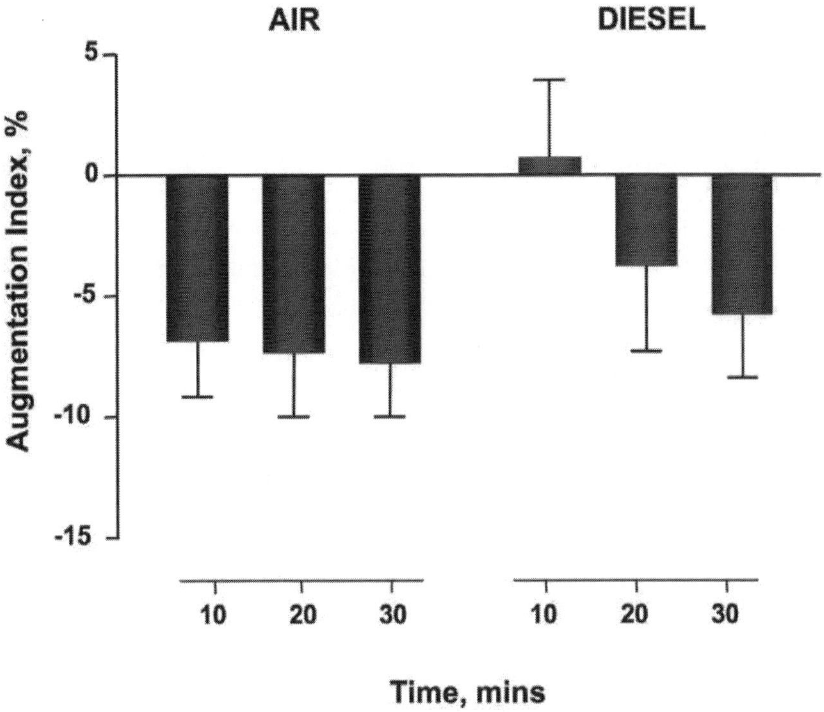

Figure 2. Augmentation index increases immediately following exposure to dilute diesel exhaust as compared to filtered air and is normalised over 30 minutes (P = 0.02 respectively, diesel exhaust ***versus*** filtered air, repeated measures ANOVA, n = 12).

Diesel exhaust exposure did not affect heart rate or brachial diastolic and systolic blood pressure during exposure (data not shown) or throughout the study period (Table 1). Augmentation pressure (AP) and augmentation index (AIx) were generally low, reflecting compliant arteries in these young healthy volunteers [8]. AP and AIx increased immediately after diesel exhaust exposure compared to filtered air (p = 0.01 and p = 0.02 respectively) and normalised over 30 minutes (Figure 1 and figure2). Time to wave reflection was reduced at 10 and 20 minutes (Table 1) consistent with an early increase in pulse wave velocity. However, formal measurement of carotid-femoral PWV 40 minutes following exposure was not different between exposures (5.1 ± 0.2 m/s versus 4.6 ± 0.1 m/s, P = ns).

DISCUSSION

Brief acute exposure to diesel exhaust is associated with a transient increase in arterial stiffness in healthy individuals. These changes in arterial stiffness occurred in the absence of any effect on peripheral arterial pressure. If our findings were confirmed in larger cohorts of susceptible populations, this simple non-invasive method of assessing arterial stiffness may become a useful technique in measuring the impact of real world exposures to combustion-derived air pollution.

As expected in our healthy young volunteers, blood pressure was in the normal range and there was no evidence of increased arterial stiffness following filtered air exposure. In contrast, exposure to diesel exhaust increased the augmentation pressure by 3 mmHg and augmentation index by 8%. Whilst these small changes are unlikely to be of significance in healthy individuals with normal cardiovascular function, these changes would be clinically important in susceptible individuals. Small increases in arterial stiffness can result in augmentation of central aortic systolic pressures that in turn increases left ventricular afterload and reduces diastolic coronary filling. In a patient with established hypertension or coronary artery disease, small changes in central aortic pressure may be sufficient to trigger an acute cardiovascular event. It is interesting to note that the effect of DE on augmentation index is of a similar magnitude to that of smoking a single cigarette [13].

To our knowledge, this is the first study to assess the effects of diesel exhaust air pollution on arterial stiffness. Our findings are consistent with those of Brook et al where a 2 hour exposure to concentrated ambient particles (CAPs) and ozone resulted in acute vasoconstriction in healthy volunteers [17]. It is entirely conceivable that this effect of CAPs on conduit vessel function would be associated with increased arterial stiffness and augmented central arterial pressures. Of note, Brook and colleagues found that exposure to CAPs caused a small but consistent increase in diastolic blood pressure by 2 mmHg [18]. Whilst changes in arterial stiffness in our cohort were not associated with an increase in blood pressure, this may be due to our shorter exposure, differences in particle composition between exposures, or the absence of any appreciable quantities of ozone in DE. Moreover, a recent field study revealed increased arterial stiffness following exposure to welding fumes [19]. Taken together, these studies underpin the findings in the present study.

A number of hypotheses have been presented to explain the mechanisms responsible for the adverse effects of PM on endothelial function and vasomotor tone [20]. Inhalation of PM causes oxidative stress in the airways and induces inflammatory response with upregulation of oxidative stress-sensitive pathways [20, 21]. Oxidative stress triggers an acute inflammatory reaction locally in the airways with the potential to upregulate systemic inflammatory processes. Acute systemic inflammation has adverse effects on endothelial vasomotor function and increases arterial stiffness [21]. Whilst we did not measure oxidative stress or systemic inflammation in our current study, we have previously demonstrated that a similar DE exposure modifies airway antioxidants [15] and increases interleukin-6 and tumour necrosis factor-alpha concentrations in plasma [22]. Vascular tone is intimately regulated by the balance between reactive oxidant species and nitric oxide. Interestingly, there are reports in preclinical studies of nanoparticle translocation from the airways into the blood stream [23, 24] with the possibility of a direct oxidative effect of PM on vascular function. Alternatively, oxidative stress may cause acute release of endothelin-1 from either pulmonary or systemic vascular endothelium [25]. Endothelin-1 is a potent vasoconstrictor that has been associated with both hypertension [26] and arterial stiffness [27] and further research into the impact of DE exposure on the endothelin system is anticipated.

Pulse wave analysis to measure arterial stiffness is a simple inexpensive and non-invasive bedside technique that allows repeated measures to determine the kinetics of air pollution-induced vascular dysfunction. Where plethysmography provides detailed characterisation of vascular mechanisms, it is time consuming, invasive, and cannot be used to assess acute or transient effects of exposure on vascular function. It is also not suitable for population-based studies.

Our studies to date have been based on carefully controlled exposures to diesel exhaust to avoid the potential for confounding that exists in observational studies. We believe this model is highly relevant both in the composition and magnitude of exposure for the assessment of short-term health effects in man. PM concentrations can regularly reach levels of 300 μg/m^3 and above in heavy traffic, occupational settings, and in the world's largest cities [28]. Exposure to 300 μg/m^3 for one hour increases a person's average exposure over a 24-hour period by only 12 μg/m^3. Changes of this magnitude occur on a daily basis even in the least polluted cities, and are associated with increases in cardiorespiratory mortality [29].

Whilst we report transient changes in arterial stiffness as indicated by an increase in AP, AIx and a reduction in the time to wave reflection, we did not find any effect of exposure on PWV. This is perhaps not surprising as we measured PWV at 40 min; a time point where AP and AIx had returned to baseline values. Whilst the early reduction in time to wave reflection would suggest an immediate increase in pulse wave velocity, time to wave reflection is a derived parameter and only a surrogate marker of PWV. Carotid-femoral PWV is considered the gold-standard measurement of central arterial stiffness [30]. In subsequent studies, it will therefore be important to assess PWV immediately post exposure to validate our findings. Given the transient nature of the change detected by pulse wave analysis, it may be necessary to measure arterial stiffness and blood pressure during the exposure itself although this will undoubtedly present technical challenges. When exploring a novel mechanism it is important to first test the hypothesis in healthy volunteers to avoid any potential confounding by disease states and medications. However, as discussed, it is unlikely that small transient changes in arterial stiffness are of clinical relevance in our cohort. Therefore, subsequent studies would be of importance to address whether this effect of diesel exhaust on arterial stiffness is also applicable to women older people, and whether the effects would be even more pronounced in susceptible patient groups with pre-existing vascular dysfunction or hypertension.

CONCLUSION

Brief exposure to diesel exhaust is associated with a transient increase in arterial stiffness in healthy individuals. This suggests that exposure to diesel exhaust impairs arterial function and provides insight into the association between air pollution and increased cardiovascular morbidity. If our findings are confirmed in larger cohorts of susceptible populations, this simple non-invasive method of assessing arterial stiffness may become a useful technique in measuring the impact of real world exposures to combustion-derived air pollution.

ACKNOWLEDGEMENTS

This work was supported by British Heart Foundation Project (PG/04/131) and Programme (PG/05/003) Grants; a Small Project grant from the National Health Service Research and Development Fund; the Swedish Heart-Lung Foundation; the Swedish Research Council for Environment, Agricultural Sciences, and Spatial Planning (FORMAS); Front Research; the County Council of Västerbotten; and the Swedish National Air Pollution Programme. We thank Frida Holmström, Annika Johansson, Maj-Cari Ledin, Jamshid Pourazar, and all the staff in the Departments of Respiratory Medicine/Allergy and the Clinical Research Facility, University Hospital, Umeå; Finny Patterson, Laura Flint and all the staff at the Wellcome Trust Clinical Research Facility, Edinburgh. Dr Blomberg is the holder of the distinguished Lars Werkö Research fellowship from the Swedish Heart-Lung Foundation.

AUTHORS' CONTRIBUTIONS

AB, NM, KD, TS and DN coordinated and were responsible for the planning of the study. ML, AL and SB were responsible for the diesel exhaust exposures. ML carried out the arterial stiffness measurements. ML and NM analysed data and performed statistical analysis. The manuscript was written by ML, NM and AB and then read, corrected and approved by all authors.

REFERENCES

1. Dockery DW, Pope CA, Xu XP, Spengler JD, Ware JH, Fay ME, Ferris BG, Speizer FE: **An association between air-pollution and mortality in 6 united-states cities.** *New England Journal Of Medicine* 1993, **329:** 1753–1759.

2. Laden F, Neas LM, Dockery DW, Schwartz J: **Association of fine particulate matter from different sources with daily mortality in six U.S. cities.** *EnvironHealth Perspect* 2000, **108:** 941–947.

3. Charron A, Harrison RM: **Fine (PM2.5) and Coarse (PM2.5–10) Particulate Matter on A Heavily Trafficked London Highway: Sources and Processes.** *Environ Sci Technol* 2005, **39:** 7768–7776.

4. Peters A, Dockery DW, Muller JE, Mittleman MA: **Increased particulate air pollution and the triggering of myocardial infarction.** *Circulation* 2001, **103:** 2810–2815.

5. Mills NL, Tornqvist H, Gonzalez MC, Vink E, Robinson SD, Soderberg S, Boon NA, Donaldson K, Sandstrom T, Blomberg A, Newby DE: **Ischemic and Thrombotic Effects of Dilute Diesel-Exhaust Inhalation in Men with Coronary Heart Disease.** *N Engl J Med* 2007, **357:** 1075–1082.

6. Brook RD, Brook JR, Urch B, Vincent R, Rajagopalan S, Silverman F: **Inhalation of Fine Particulate Air Pollution and Ozone Causes Acute Arterial Vasoconstriction in Healthy Adults.** *Circulation* 2002, **105:**1534–1536.

7. Mills NL, Tornqvist H, Robinson SD, Gonzalez M, Darnley K, MacNee W, Boon NA, Donaldson K, Blomberg A, Sandstrom T, Newby DE: **Diesel Exhaust Inhalation Causes Vascular Dysfunction and Impaired Endogenous Fibrinolysis.** *Circulation* 2005, **112:** 3930–3936.

8. McEniery CM, Yasmin , Hall IR, Qasem A, Wilkinson IB, Cockcroft JR: **Normal Vascular Aging: Differential Effects on Wave Reflection and Aortic Pulse Wave Velocity: The Anglo-Cardiff Collaborative Trial (ACCT).***Journal of the American College of Cardiology* 2005, **46:** 1753–1760.

9. Oliver JJ, Webb DJ: **Noninvasive Assessment of Arterial Stiffness and Risk of Atherosclerotic Events.***Arterioscler Thromb Vasc Biol* 2003, **23:** 554–566.

10. Laurent S, Boutouyrie P, Asmar R, Gautier I, Laloux B, Guize L, Ducimetiere P, Benetos A: **Aortic Stiffness Is an Independent Predictor of All-Cause and Cardiovascular Mortality in Hypertensive Patients.***Hypertension* 2001, **37:** 1236–1241.

11. McEniery CM, Wallace S, Mackenzie IS, McDonnell B, Yasmin , Newby DE, Cockcroft JR, Wilkinson IB:**Endothelial Function Is Associated With Pulse Pressure, Pulse Wave Velocity, and Augmentation Index in Healthy Humans.** *Hypertension* 2006, **48:** 602–608.

12. Pauca AL, O'Rourke MF, Kon ND: **Prospective Evaluation of a Method for Estimating Ascending Aortic Pressure From the**

Radial Artery Pressure Waveform. *Hypertension* 2001, **38:** 932–937.

13. Mahmud A, Feely J: **Effect of Smoking on Arterial Stiffness and Pulse Pressure Amplification.** *Hypertension*2003, **41:** 183–187.

14. Lucking AJ, Lundback M, Mills NL, Faratian D, Barath SL, Pourazar J, Cassee FR, Donaldson K, Boon NA, Badimon JJ, *et al.*: **Diesel exhaust inhalation increases thrombus formation in man.** *Eur Heart J*2008,**29**(24):3043–3051.

15. Behndig AF, Mudway IS, Brown JL, Stenfors N, Helleday R, Duggan ST, Wilson SJ, Boman C, Cassee FR, Frew AJ,*et al.*: **Airway antioxidant and inflammatory responses to diesel exhaust exposure in healthy humans.***European Respiratory Journal* 2006, **27:** 359–365.

16. Mills NL, Miller JJ, Anand A, Robinson SD, Frazer GA, Anderson D, Breen L, Wilkinson IB, McEniery CM, Donaldson K, *et al.*: **Increased arterial stiffness in patients with chronic obstructive pulmonary disease: a mechanism for increased cardiovascular risk.** *Thorax* 2008, **63:** 306–311.

17. Brook Rd BJRU: **Inhalation of fine particulate air pollution and ozone causes acute arterial vasoconstriction in healthy adults.** *Circulation* 2002,**105**(13):1534–6.

18. Urch B, Silverman F, Corey P, Brook JR, Lukic KZ, Rajagopalan S, Brook RD: **Acute blood pressure responses in healthy adults during controlled air pollution exposures.** *Environ Health Perspect* 2005,**113**(8):1052–1055.

19. Fang SEE, Cavallari J, Mittleman M, Christiani D: **Acute changes in vascular function among welders exposed to metal-rich particulate matter.** *Epidemiology* 2008,**19**(2):217–225.

20. Brook RD, Franklin B, Cascio W, Hong Y, Howard G, Lipsett M, Luepker R, Mittleman M, Samet J, Smith SC Jr, Tager I: **Air Pollution and Cardiovascular Disease: A Statement for Healthcare Professionals From the Expert Panel on Population and Prevention Science of the American Heart Association.** *Circulation* 2004,**109:** 2655–2671.

21. Hingorani AD, Cross J, Kharbanda RK, Mullen MJ, Bhagat K, Taylor M, Donald AE, Palacios M, Griffin GE, Deanfield JE, *et*

al.: **Acute Systemic Inflammation Impairs Endothelium-Dependent Dilatation in Humans.** *Circulation* 2000, **102:** 994–999.

22. Tornqvist H, Mills NL, Gonzalez M, Miller MR, Robinson SD, Megson IL, MacNee W, Donaldson K, Soderberg S, Newby DE, *et al.*: **Persistent Endothelial Dysfunction in Humans after Diesel Exhaust Inhalation.** *Am J Respir Crit Care Med* 2007, **176:** 395–400.

23. Oberdorster G, Sharp Z, Atudorei V, Elder A, Gelein R, Lunts A, Kreyling W, Cox C: **Extrapulmonary translocation of ultrafine carbon particles following whole-body inhalation exposure of rats.** *J ToxicolEnviron Health A* 2002, **65:** 1531–1543.

24. Nemmar A, Hoylaerts MF, Hoet PH, Dinsdale D, Smith T, Xu H, Vermylen J, Nemery B: **Ultrafine particles affect experimental thrombosis in an in vivo hamster model.** *AmJRespirCrit Care Med* 2002, **166:** 998–1004.

25. Bouthillier L, Vincent R, Goegan P, Adamson IYR, Bjarnason S, Stewart M, Guenette J, Potvin M, Kumarathasan P: **Acute Effects of Inhaled Urban Particles and Ozone: Lung Morphology, Macrophage Activity, and Plasma Endothelin-1.** *Am J Pathol* 1998, **153:** 1873–1884.

26. Schiffrin EL: **Role of Endothelin-1 in Hypertension.** *Hypertension* 1999, **34:** 876–881.

27. McEniery CM, Qasem A, Schmitt M, Avolio AP, Cockcroft JR, Wilkinson IB: **Endothelin-1 regulates arterial pulse wave velocity in vivo.** *Journal of the American College of Cardiology* 2003, **42:** 1975–1981.

28. World Health O: *Urban air pollution in megacities of the world.* Blackwell; 1992.

29. Samet JM, Dominici F, Curriero FC, Coursac I, Zeger SL: **Fine particulate air pollution and mortality in 20 U.S. cities, 1987–1994.** *NEnglJ Med* 2000, **343:** 1742–1749.

30. O'Rourke MF, Franklin SS: **Arterial stiffness: reflections on the arterial pulse.** *Eur Heart J* 2006, **27:** 2497–2498.

Chapter 9

Assessment of Diesel Particulate Matter Exposure Among Underground Mine Workers

D. Bertolatti, K. Rumchev & B. Mullins

Curtin Health Innovation Research Institute and The School of Public Health, Curtin University, Australia

ABSTRACT

Diesel motor emissions are a major source of ultrafine particles as research shows that the particulate fraction of diesel exhaust consists mainly of very small particles, which rapidly agglomerate to form clumps of particles with <1 μm aerodynamic size. Substantial scientific evidence supports that inhalation of emissions from diesel exhaust particulates is associated with a range of adverse health effects. Measurements have shown that underground miners can be exposed to over 100 times the typical environmental concentrations of diesel particulate matter (DPM) and over 10 times the concentration measured in other work environments where diesel engines are common. This paper reports personal and environmental exposures to DPM in an underground mine in Western Australia. The findings demonstrated that the DPM concentration exceeded the provisional 8-hour time weighted average exposure standard of 0.1 mg/m3 for most of the assessed locations. The elemental analysis indicated high concentrations of silica and iron in DPM which could be potentially associated with some adverse health effects including respiratory illnesses and irritating effects. Thermal comfort underground was also assessed and the results showed non compliance with the standards of the American Society of Heating, Refrigerating and Air-Conditioning Engineers (ASHRAE).

Keywords: diesel exhaust, diesel particulate matter, fine and ultrafine particles.

1 INTRODUCTION

Exhaust particles emitted from in-use diesel vehicles in underground mining operations are a major source of ultrafine carbonaceous particles, which have the potential to produce adverse health effects. The particulate fraction of diesel exhaust consists of fine particles (typically aerodynamic diameter < 2.5 µm) including a high number of ultrafine particles (diameter < 0.1 µm) that include reactive organic and transition metal components [1].

Particles in the ultrafine size range have attracted significant scientific and medical attention since research shows that for equivalent masses of inhaled particles, ultrafine particles provide a greater surface area for adsorption of potentially toxic substances and metal agents. Further to this, ultrafine particles have also a higher deposition probability particularly in small airways and the alveolar region of the lungs than coarse particles. With regards to health outcome, the small particle size of diesel particulate matter (DPM) is considered a significant health issue as such particles can penetrate deep in the lungs [2–4]. In 1988, the National Institute of Occupational Safety and Health (NIOSH) [5] recommended that diesel exhaust be regarded as a "potential occupational carcinogen". Similarly, in 1989, the International Agency for Research on cancer declared that "diesel engine exhaust is probably carcinogenic to human" [6]. Measurements have shown that underground miners can be exposed to over 100 times the typical environmental concentrations of diesel particulate matter (DPM) and over 10 times the concentration measured in other work environments where diesel engines are common [7–9].

Based on the current research evidence, the US National Research Council has concluded that the NIOSH Research Mining Program, related to exposure of DPM and related health effects, is considered as a high priority with a rating of four on a five-point scale [10].

Furthermore, evidence is increasingly being accumulated which suggests that nanoparticles (such as diesel soot which typically has a primary particle size <50 nm) affect health as a function of surface area rather than mass [11].

The study assessed personal and environmental exposures of diesel particulate matter (DPM) in an underground mine using various sampling system that included, low pressure cascade impactor (Nano-MOUDI-II) and scanning mobility particle size (SMPS) measurements. This research forms part of a wider study with the major milestone to reduce exposure levels of diesel particulate matter (DPM) and improve the air quality underground.

2 METHODS AND MATERIALS

The study was undertaken in a large underground mine in north-western Australia. The experimental method was designed to: (a) accurately measure DPM in real-time at representative locations, (b) collect size-fractioned DPM and other particulate for gravimetric measurement and analysis, (c) determine worker exposure levels.

2.1 Apparatus

The experimental apparatus consisted of, a NanoMOUDI 125B multi-stage low pressure impactor (MSP, Shoreview, MN, USA), a Scanning Mobility Particle Sizer (SMPS) consisting of a TSI 3081 Differential Mobility Analyzer (DMA) and TSI 3775 Condensation Particle Counter (CPC) (TSI, Shoreview, MN, USA), a TSI DustTrak 8533 DRX, with a TSI P-Trak, TSI Velocicalc-Plus used to obtain air velocity, temperature and humidity measurements. Four SKC air pumps and filters were utilized for personal monitoring of mine-worker exposures. NIOSH method 5040 (Elemental Carbon) was applied to quantify elemental carbon (EC) via thermal-optical analyzer. Samples collected using the NanoMOUDI were analysed for elemental composition of by Scanning Electron Microscopy (SEM) Analysis and performed by the ChemCentre of WA. Analysis of organic species and PAHs was also undertaken.

2.1.1 Methods

Measurement sites were established at 4 representative locations within the mine:

Site 1 – The main fresh air inlet (baseline measurement) for the mine.
Site 2 – A dead-end passage at the base of the tunnel network with no airflow.

Site 3 – An active area of the mine where material extraction was actively occurring.

Site 4 – A site near the main air outlet from the mine.

Figure 1. Sampling equipment located on mine vehicle, in location at site 1.

Samples were taken at a height equivalent to the roof level of the vehicle. Sampling commenced at least 30 minutes after the vehicle was placed into position. The NanoMOUDI is not visible in image. Measurements were also taken above ground to compare with the fresh air inlet values.

At each site 1-4, measurements were taken continuously for a period of 6– 14 hours. During each measurement, the NanoMOUDI, DustTrak and P-Trak were set to measure continuously. The SMPS was set to perform repeated scans between 15 nm and 950 nm at intervals of approximately 15 minutes. At the end of each measurement the equipment was taken to a clean laboratory on the surface where samples/filters could be removed and weighed, and data downloaded, etc. At the start of each working shift, three workers were fitted with personal samplers, selected based on their proximity to the sampling location. The samplers were collected after 8 hours and the filters removed for analysis. For the time period where samples were being collected, information on vehicle movements and vehicle locations were obtained from the mine telemetry/logistics system. This made it possible to track vehicle movement past the sampling location.

3 RESULTS AND DISCUSSION

Table 1 shows the airflow, temperature and humidity values recorded at each sampling location.

Table 1. Environmental conditions at each sampling location.

Site	Location	Mean Temp (^{0}C)	Mean R.H. (%)	Mean Air Vel. (m/sec)
1	Fresh air inlet	11.7	75%	2.01
2	Dead end tunnel	36.5	98%	0.17
3	Active extraction	30.2	83%	0.18
4	Near air exhaust	31.1	90%	0.57
Recommended range (ASHRAE 2004)		21–26	30–70	0.1–0.2

It will be noted that humidity levels are extremely high. This was due to a combination of evaporative cooling of the mine intake air, water sprays for dust suppression within the mine, and also artesian water which enters the mine and evaporates. It will be noted that relative humidity and temperature in the dead end tunnel were both exceptionally high. During the time of the measurement campaign, the mine was operating at approximately 30% of its typical operating capacity, due to the global financial crisis reducing demand/value of the mine production. Therefore particulate levels measured during this campaign could be expected to be near the lower levels of typical worker exposure.

3.1 Personal monitoring

Table 2 shows the results for personal monitoring, for sites 1, 3 and 4, averaged over 8 hours. Site 2 did not have any workers in the area while sampling was conducted, hence no data was possible.

The samplers were fitted to workers at the start of each shift. Workers would typically spend the first hour or more of each shift above ground, and then may come above ground for lunch and other breaks (though this is not enforced and underground refuge areas are often used for breaks). The vehicle operators have air conditioned vehicle cabins; however some choose to operate their vehicles with the windows open. The drill vehicles are the only electric vehicles in the mine, which accounts for the low exposure of the driller and assistant driller. The underground service/maintenance mechanics and the explosives technicians are the only workers who are usually required to spend extended periods out of a vehicle in the mine, which accounts for the elevated readings for these workers. However these latter two types of workers drive between tasks in air conditioned 4x4 vehicles.

Table 2. Personal exposure to DPM (measured as elemental carbon).

Site	Worker	Elemental Carbon (mg/m^3)
1	Loader ("Bogger") driver	0.06
	Truck driver	0.03
	Fixed position	<0.002
3	Assistant driller	0.03
	Drill vehicle driver	0.03
	Fixed position	0.05
4	Explosives technician	0.078
	Service/breakdown mechanic	0.11
	Fixed position	0.085

3.1.1 Stationary sampling results

Table 3 below shows the PM 1 levels measured by NanoMOUDI. This was selected as it is the size range which includes almost all DPM. EC values have also been determined from the total gravimetric measurements of PM 1. The aerosol in the mine was found to be bimodal. There existed a sub-micron peak, which consisted almost entirely (>99% by mass) of DPM, and a super-micron peak (at approximately 2-5 microns), which consisted of crustal materials from the mine, combined with high levels of silica (likely from the spray-concrete used to secure the ceiling and walls of the mine.

Table 3. PM1 and DPM levels as measured by NanoMOUDI.

Site	Sampling Time (hours: mins)	Average PM1 conc. (mg/m^3)	EC (mg/m^3)
1	6:00	0.08 mg/m^3*	0.03 mg/m^3*
2	8:50	0.11 mg/m^3	0.09 mg/m^3
3	6:55	0.16 mg/m^3	0.13 mg/m^3
4	14:10	0.12 mg/m^3	0.10 mg/m^3

*A high level of elemental carbon was present in the intake air due to wildfires in the vicinity of the mine preceding the sampling period.

It can be observed that a nontrivial proportion of the (clean) intake air was elemental carbon. This was believed to largely be due to recent wildfires, rather than vehicle movement on the surface. It will be noted that the average levels are very high, despite the fact that the mine was operating at approximately 30% of normal production levels.

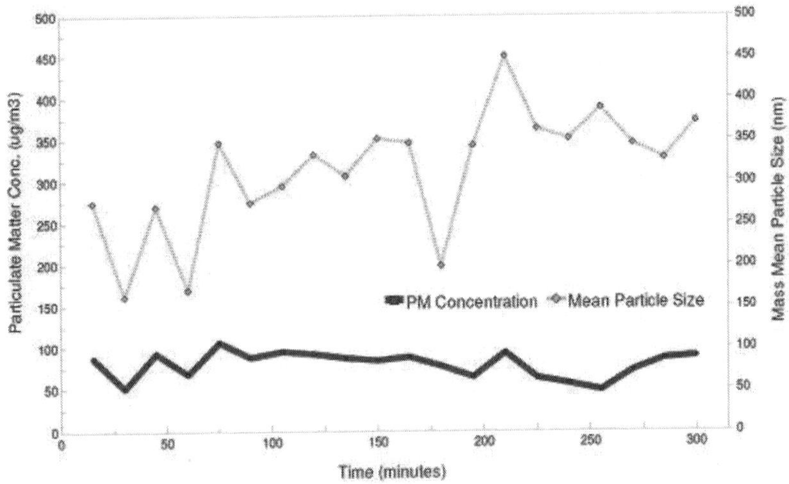

(a)

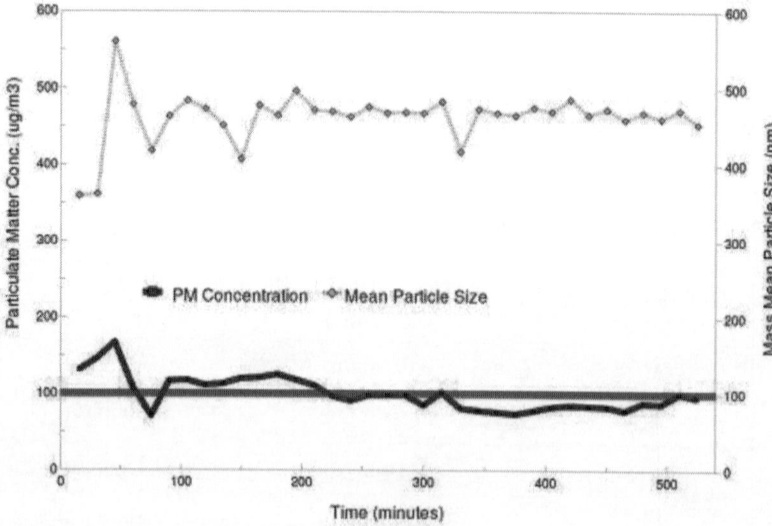

(b)

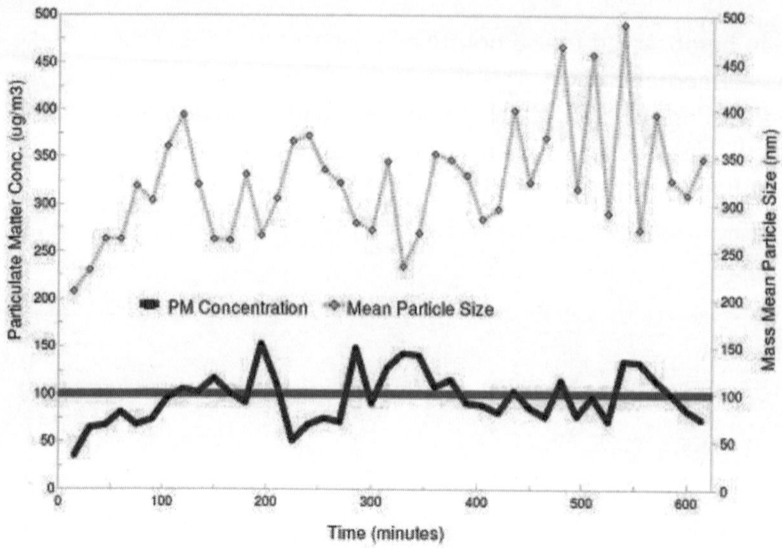

(c)

Figure 2. Time dependent mass and size of PM 1 (DPM) as measured using the SMPS. (a) Site 1, (b) Site 2, (c) Site 3. The red line denotes the time weighted DPM exposure limit of 0.1 mg/m^3.

Figure 2(a–c) shows the temporal evolution of PM 1 at sites 1–3. It can be observed that at sites 1 and 2 the values were relatively stable, as site 1 has a constant airflow of air which has passed through a filtration and conditioning system, and Site 2 has a low airflow and a very low air exchange rate. Site 3 however, had a mine vehicle (loader/bogger) working in close proximity to the sampling apparatus. It will be observed than an increase in particle mass often corresponds to a decrease in particle size. The mine telemetry data showed that these samples coincided with the loader passing by the sampling location, resulting in a high concentration of smaller particles which had just been emitted from the exhaust, and had not had sufficient time to dissipate or agglomerate.

3.1.2 Elemental composition of PM

An elemental analysis of the PM collected by the NanoMOUDI was also performed. The total particulate matter values for all size classes have been combined. Over 50% of the inorganic contaminants were found in the PM 2.5 fraction.

It can be observed that high levels of Silica and Iron were present, with levels of vanadium also elevated. The Silica is likely due to the concrete which is sprayed onto the walls and ceiling of the mine to ensure structural integrity. Overall, both the DPM and inorganic levels in the mine particulate are of concern. Overall TSP and PM10 levels were not significant, as the mass of larger particles is kept low due to water sprays, gravitational settling and the humidity in the mine.

In general, both the DustTrak and the SMPS showed a reasonably good agreement with each other and with the SMPS, generally within 10%. However both the DustTrak and especially the SMPS showed elevated mass levels at sites 2 and 4, likely due to condensed water on the aerosol particles. This was not a problem for NanoMOUDI samples were weighed and analysed in a dry state.

Direct reading instruments such as the DustTrak should always be calibrated against a gravimetric standard. For the mine in question, it would be advisable to fit a diffusion dryer before the sample inlet, so as to remove excess moisture, allowing the dry particle mass to be measured. The work showed that a P-Trak is unsuitable for DPM measurement as DPM mass was often inversely proportional to number concentration.

Table 4. Inorganics found in PM10. Site 1–4 values are averages over the sampling period in mg/m^3.

Element	TWA (NIOSH)	Site 1	Site 2	Site 3	Site 4
Magnesium (mg/m^3)	10 mg/m^3	0.002	0.005	0.002	0.005
Nickel (mg/m^3)	0.015 mg/m^3	0.001	0.003	0.004	0.005
Silica, (mg/m^3)	0.05 mg/m^3	0.050	0.044	0.071	0.073
Vanadium dust (mgV/m^3)	0.05 mgV/m^3 (15 min)	0.001	0.002	0.005	0.005
Iron (mg/m^3)	0.23 mg/m^3	0.056	0.127	0.202	0.195

As mentioned, the mine was operating at only 30% of its usual capacity while the measurements were performed. It could be expected that the DPM values could triple once production is scaled back up at the mine. However, the ventilation and air conditioning system (or at least the fans which supply the air) were reportedly operating at 100% capacity. Due to the humidity, temperature and DPM problems which were found, significant remedial action is required.

4 CONCLUSION

This work provided a comprehensive evaluation of the DPM and PM levels in an underground mine, including the spatial and temporal information, as well as composition analysis.

Very high levels of DPM were found in the mine, which are likely to only increase once production levels are increased. There was however a slight disconnect between mine average DPM and worker exposure, as worker practices differ, and some workers spend most of their time in air conditioned cabs.

For the mine in question, optimisation of the ventilation system would be advisable, couples with engineering controls such as additional filtration or particulate removal/agglomeration technology. Further study into DPM levels in Australian mines and the formulation of binding exposure limits is warranted.

REFERENCES

1. Wichmann, H.E., Diesel Exhaust Particles. Inhalation Toxicology, 19(Suppl.1), pp. 241-244, 2007.
2. Mühlfeld, C., Rothen-Rutishauser, B., Blank, F., Vanhecke, D., Ochs, M. & Gehr, P., Interactions of Nanoparticles With Pulmonary Structures and Cellular Responses. American Journal of Physiology - Lung Cell Molecular Physiology, 294(5), pp. L817-L829, 2008.
3. Diesel Exhaust: A Critical Analysis of Emissions, Exposures, and Health Effects. A Special Report of the Institute's Working Group; Health Effects Institute Web Site, pp. 294, 1995. pubs.healtheffects.org/view.php?id= 1 1
4. Pope, A., Burnett, R., Thun, M., Calle, E., Krewski, D., Ito, K. & Thurston, G., Lung Cancer, Cardiopulmonary Mortality, and Long-Term Exposure to Fine Particulate Air Pollution. Journal of the American Medical Association, 287(9), pp. 1132-1141, 2002.
5. Carcinogenic Effects of Exposure to Diesel Exhaust; National Institute of Occupational Safety and Health (NIOSH), Current Intelligence Bulletin 50, 1988, Cincinnati, Ohio, www.cdc.gov/niosh/88116_50.html
6. Monographs on the Evaluation of Carcinogenic Risks to Humans Diesel and Gasoline Engine Exhausts and Some Nitroarenes; World Health Organisation, International Agency for Research on Cancer website, 46, p. 458, 1989, Lyon, France. monographs.iarc.fr/ENG/Monographs/vol46 /volume46.pdf
7. Cantrell, B.K. & Watts, W.F., Diesel Exhaust Aerosol: Review of Occupational Exposure. Applied Occupational and Environmental Hygiene, 12(12), pp 1019-1026, 1997.

8. Cash, D. & Baughman, W., DPM exposure in metal and nonmetal mines in the United States 2002-2005 and the 2005 Final Rule on the Interim Limit. Proceeding of Mining Diesel Emissions Conference (MDEC), Markham, Ontario, October 12-14, 2005.

9. Bugarski, Schnakenberg & Patts, Implementation of Diesel Particulate Filter Technology in Underground Metal and Nonmetal Mines; National Institute for National Safety and Health, Centre for Disease Control Web Site, Pittsburgh, USA, www.cdc.gov/niosh/mining/pubs/pdfs/iodpf.pdf

10. Mining Safety and Health Research at NIOSH: Reviews of Research Programs of the National Institute for Occupational Safety and Health. NISBN: 0-309-66636-8, National Academies Press: Washington DC, 2007.

11. Biswas, P. & Wu, C. Nanoparticles and the environment. Journal of the air & waste management association, 55(6), pp. 708-746, 2005.

Chapter 10

Co-Exposure to Allergen and Diesel Exhaust Enhance Inflammatory Responses in Human Airway Submucosa

Ali Hosseini[1,2*], Tillie L Hackett[2], Jeremy Hirota[1,2], Kelly McNagny[3] and Chris Carlsten[1,2]

[1]*Department of Medicine, University of British Columbia, Vancouver, British Columbia, V6T 1Z3, Canada.*

[2]*University of British Columbia, James Hogg Research Centre of the Heart + Lung Institute, Vancouver, British Columbia, V6Z 1Y6, Canada.*

[3]*University of British Columbia, Biomedical Research Centre, Vancouver, British Columbia, V6T 1Z3, Canada.*

BACKGROUND

Asthma is a chronic condition described by inflammation of the airways and lungs. Diesel exhaust (DE) is a major contributor to ambient particulate matter (PM) air pollution. There is rising evidence that PM acts as adjuvant on the immune responses and may lead to augmentation of allergic inflammation [1,2]. We aim to elucidate if DE increases allergen-induced inflammation and cellular immune response in the airways of atopic human subjects.

METHODS

15 volunteer participants with allergy to house dust mite allergen (Der p 1), birch or Timothy grass were recruited. In a randomized fashion, subjects inhaled DE (300µg $PM_{2.5}/m^3$) or filtered air for 120 minutes. One hour following the exposure, the extract of an aeroallergen to which the individual is sensitive, or placebo (sterile saline), was instilled into contralateral lung segments through bronchoscopy. Endobronchial biopsies from these same segments were then acquired 48 hours after each exposure. This was repeated 4 weeks later in each subject with the alternative inhalant. Thus, biopsies under 4 different conditions were created: filtered air + saline (FAS), DE + saline (DES), filtered air + allergen (FAA) and DE + allergen (DEA). Biopsies were processed and embedded in glycol methanlacrylate acrylic resin and serial sections were cut to 2µm and used for immunostaining with monoclonal antibodies to tryptase and eosinophil cationic protein (ECP). The percent positivity and distribution of activated mast cells (tryptase+) and eosinophils (ECP+) were quantified in the bronchial submucosa by Aperio ImageScope software.

RESULTS

The percent positivity for tryptase expression: FAS=0.54±0.05, DES=0.51±0.18, FAA=0.63±0.24, DEA= 0.94±0.23. The percent positivity for ECP expression: FAS=0.35±0.17, DES=0.38±0.11, FAA=0.61±0.14, DEA=0.73±0.33. Data are presented as mean ± SEM (n=6).

CONCLUSIONS

Our preliminary data suggest that DE may enhance the inflammatory response to allergen in atopic individuals. This data is novel in the context of human lung tissue.

ACKNOWLEDGEMENTS

This study is funded by the Canadian Institutes of Health Research (CIHR). A.H. is supported by CIHR Transplantation Scholarship Training Program.

REFERENCES

1. Riedl M, Diaz-Sanchez D: Biology of diesel exhaust effects on respiratory function. Journal of Allergy and Clinical Immunology. 2005, 115 (2): 221-228.

2. Nel AE, Diaz-Sanchez D, Ng D, Hiura T, Saxon A: Enhancement of allergic inflammation by the interaction between diesel exhaust particles and the immune system. Journal of Allergy and Clinical Immunology. 1998, 102 (4): 539-554.

Index